Pluto

Sentinel of the Outer Solar System

Orbiting at the edge of the outer Solar System, Pluto is an intriguing object in astronomy. Since the fascinating events surrounding its discovery, it has helped increase our understanding of the origin and evolution of the Solar System and has raised questions about the nature and benefits of scientific classification.

This is a timely and exciting account of Pluto and its satellites. The author uses Pluto as a case study to discuss discovery in astronomy and how remote astronomical bodies are investigated. He also examines the role of classification in science by discussing Pluto's recent classification as a dwarf planet. Besides Pluto, the book also explores the rich assortment of bodies that constitute the Edgeworth-Kuiper Belt, of which Pluto is the innermost substantial member.

Richly illustrated and up to date, this book is written for general readers, amateur astronomers and students alike. Boxed text provides more advanced information especially for readers who wish to delve deeper into the subject.

BARRIE W. JONES is Emeritus Professor of Astronomy in the Department of Physics and Astronomy, The Open University. A highly regarded university lecturer, he has an outstanding record in the public understanding of science, particularly in astronomy, through lectures, local and national radio and TV and articles in popular magazines and in the press. His main research area is the habitability of planetary systems.

Pluto

Sentinel of the Outer Solar System

BARRIE W. JONES

The Open University

CAMBRIDGE
UNIVERSITY PRESS

CAMBRIDGE UNIVERSITY PRESS
Cambridge, New York, Melbourne, Madrid, Cape Town, Singapore,
São Paolo, Delhi, Dubai, Tokyo, Mexico City

Cambridge University Press
The Edinburgh Building, Cambridge CB2 8RU, UK

Published in the United States of America by Cambridge University Press,
New York

www.cambridge.org
Information on this title: www.cambridge.org/9780521194365

First published 2010

Printed in the United Kingdom at the University Press, Cambridge

A catalogue record for this publication is available from the British Library

Library of Congress Cataloguing in Publication data
Jones, Barrie William.
Pluto : sentinel of the outer solar system / Barrie W. Jones.
   p.   cm.
Includes bibliographical references and index.
ISBN 978-0-521-19436-5 (hardback)
1. Pluto (Dwarf planet)   2. Solar system – Origin.   3. Kuiper Belt.   I. Title.
QB701.J66   2010
523.49'22 – dc22   2010015480

ISBN 978-0-521-19436-5 Hardback

To my wife Anne, to all other members of my family, and in memory of my parents

Contents

Tables

The colour plates are between pages 148 and 149.

Preface

Pluto is a very tiny, distant world. It orbits the Sun beyond the giant planet Neptune, the outermost of the other eight planets in the Solar System. In inward order from Neptune these planets are Uranus, Saturn, Jupiter, Mars, Earth, Venus and Mercury. Pluto has a diameter a little less than one fifth of the diameter of our planet, which itself is a long way from being the largest planet in the Solar System. That title belongs to the giant planet Jupiter, with a diameter just over 11 times that of the Earth.

Why should a book be devoted to such a tiddler among the planets? There are three main reasons. First, the discovery of Pluto in 1930 is a fascinating episode in our quest to discover whether the Solar System beyond Neptune is devoid of planetary bodies. Second, ever since its discovery, controversy has been rampant about what sort of body Pluto is. Is it deserving of the status of planet, or is it too small for that? The classification of Pluto is an excellent example of the role of classification in all branches of science: classification not only comes with great advantages but also with difficulties. Third, Pluto is the closest large member of the Edgeworth-Kuiper belt, a great swarm of small bodies that orbit the Sun beyond Neptune. Though the existence of such a belt had been predicted in the 1940s, it was not until the 1990s that discoveries of other trans-Neptunian bodies were made. By learning about Pluto we learn something about the more distant bodies, and therefore learn more about how the Solar System formed and evolved.

A valuable spin-off is that in describing how Pluto has been explored, and will be explored, you will meet techniques in astronomy of wide applicability, not only to other bodies in the Solar System but, in a few instances, to much larger bodies, namely, the stars and to any planets that they might possess. In this respect, Pluto is a case study, but none the worse for that.

I have aimed this book at a wide readership, indeed at anyone interested in the outer reaches of the Solar System and also in tales of discovery and in how we learn about objects far away. I have assumed that you bring to this book no knowledge of astronomy, and almost no mathematics – only a basic ability in arithmetic is required and only in a few places.

To meet the needs of those able and wishing to go deeper I have used boxed text. Nothing that follows afterwards requires that you have studied the text. If you are not familiar with the contents of such boxes, you can read the text to improve your overall understanding. Boxed text is also used to separate from the main story material that would interrupt the flow, and this material is at the general level of the main text. Each box is tagged so that you know what sort it is.

Under *Further reading and other resources* I've listed some of the key papers in the scientific literature. These are to enable you to read about the various topics in this book in greater detail or greater depth. Most of these require some knowledge of astronomy and mathematics. I've also listed a few books about the Solar System, Pluto and the outer Solar System that, like this book, are aimed at a wide readership; other books listed require more background in astronomy and mathematics, and are labelled accordingly. Finally, I've given details of a few magazines and internet links.

Throughout the book you will see how our knowledge and understanding of Pluto and the outer Solar System has changed through the decades. You will also see that many uncertainties remain, and much is unknown. This is the very essence of science, a story that will never be finished, never be finalized. I wish you an enjoyable and informative read.

Acknowledgements

I am grateful to Nick Sleep for commenting on a draft of the whole book, and to William Grundy, David Jewitt, Pedro Lacerda and Alan Stern for commenting on individual chapters. Marc Guie, Dale Cruikshank, James Elliot, Amanda Gulbis, Charles Harding, Jonathan Horner, Brian Marsden, William McKinnon, Olivier Mousis, Jay Pasachoff and Judy Pipher have supplied information and clarification in response to my requests. Peter Hingley has assisted with completing several references to key papers in the scientific literature.

Illustrations have come from a variety of sources, acknowledged wherever possible in the captions. Antoinette Beiser, Christine Colburn, Mark Hurn, Debbie James and Sandrine Marchal have helped me find several photographs.

Simon Mitton was instrumental in getting my proposal for this book accepted by Cambridge University Press, and Vince Higgs and his team at CUP have helped to get it into production.

I The Solar System

Though Pluto, and the far-flung depths of the Solar System, is the focus of this book, it is essential that Pluto is placed in the context of the planetary system that it inhabits – our Solar System. In the first place, this is because Pluto is just one of a large and varied number of bodies that orbit the Sun, and cannot be treated as an isolated body in space. Secondly, much of the material in this chapter is needed to support and enhance your understanding of subsequent chapters.

But before we get to the Solar System, I start by examining its cosmic neighbourhood: a vast assemblage of stars called the Galaxy, which we see in the sky as the Milky Way.

1.1 A JOURNEY INTO OUR GALAXY

The Sun, which is at the centre of the Solar System, is one of about two hundred thousand million stars that make up the Galaxy. From extensive observations made from Earth it is clear that it has a beautiful form that, face-on, is something like that in Figure 1.1.

The stars, of various kinds, plus tenuous interstellar gas and dust, often woven into stunning forms, are concentrated into a disc highlighted by spiral arms (Figure 1.1). In our Galaxy the disc is about 100 000 light years in diameter (see Box 1.1), and most stars are in a thin sheet about 1000 light years thick – roughly the same ratio of diameter to thickness as a CD. This sheet is called the thin disc. It is enclosed in what is called the thick disc, which is about 4000 light years thick, where the space density of stars is less. The spiral arms are delineated by a high space density of particularly luminous stars and luminous interstellar clouds. Elsewhere in the disc the space density of the stars and interstellar clouds is no less; it is just that they are not as bright. At its centre the disc has a bulge called the nuclear bulge,

FIGURE 1.1 A face-on view of a spiral galaxy rather like ours. This has the galactic catalogue number NGC1232. (European Southern Observatory)

also full of stars and interstellar matter. It is very roughly 10 000 light years across. The bulge is visible as the bright central region of the galaxy in Figure 1.1; it is not quite spherical but slightly flattened. As in our Galaxy it is also slightly elongated in one direction in the plane of the disc. The disc is enveloped in the halo (not visible), a roughly spherical volume in which interstellar matter is particularly tenuous and the space density of stars is low. Throughout the Galaxy there are many groupings of stars, from binaries (two stars orbiting each other) to a variety of much larger groupings, but the Sun is an isolated star.

The Sun is located near the edge of a spiral arm, roughly half way from the centre of the Galaxy to the edge of the disc. Figure 1.2 shows the view we get from Earth of the disc of our Galaxy. This

BOX I.I THE LIGHT YEAR (PLEASE READ)
This is a unit of distance, *not time*. It is the distance that light travels in a vacuum in one year. The speed of light in a vacuum is 299 792.458 kilometres per second (1 kilometre = 0.621371 miles). In a year light travels 9.460536×10^{12} kilometres (10^{12} is 1 000 000 000 000, i.e. 1 followed by 12 zeroes). With space being near enough a vacuum, this immense unit is appropriate for expressing distances in the Galaxy. It is also appropriate for expressing interstellar distances: the nearest star to the Sun, Proxima Centauri, is 4.22 light years from the Sun. However, the Solar System is small compared with interstellar distances – the Sun is 0.0000158 light years from the Earth, which is 8.317 light minutes. The light minute would be an appropriate unit of distance within the Solar System, but as you will see in Section 1.2, a different unit is used instead.

shows part of the Milky Way, and so our Galaxy is often called the Milky Way Galaxy.

Beyond our Galaxy there are many more, some with a spiral form like ours, but there are other configurations too; some have highly irregular forms, others lack any concentration of stars and interstellar matter into a disc. It is estimated that there are tens of billions of galaxies that could be seen by our present telescopes (a billion is a thousand million).

Let's return to our Galaxy, and to that location near the edge of a spiral arm, roughly half way from the centre of the Galaxy to the edge of the disc, where the Solar System resides.

I.2 THE SOLAR SYSTEM: SIZES AND ORBITS

Sizes

The Solar System consists of a variety of bodies orbiting the Sun, plus a variety of natural satellites ('moons') orbiting most of the planets. Figure 1.3 shows the radii of bodies in the Solar System that are large enough to be spherical, and also have well known radii. You

FIGURE 1.2 The Milky Way – our view of the disc of our Galaxy.
(Naoyuki Kurita, by permission) (See plate section for colour version.)

can see, for example, that the Sun is nearly 10 times the radius of
the largest planet Jupiter. This means that it has nearly $10 \times 10 \times
10 = 1000$ times Jupiter's volume. When comparing bodies, relative
volumes give a better impression of relative sizes than relative radii.
The radius of the Earth is 6378 kilometres (km), so its diameter is

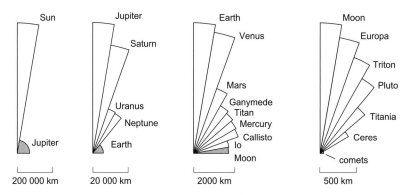

FIGURE I.3 Radii of bodies in the Solar System large enough to be spherical and with well known radii. (The relative size of comets is also indicated.) Note the scales.

twice this, 12 756 km. More precisely these are the equatorial values. The Earth's rotation around its polar axis slightly flattens it, so the radius pole to pole is 6357 km and the diameter 12 714 km. All the planets are slightly flattened by rotation, the amount depending on the rate of rotation and the composition of the planet.

Orbits

Figure 1.4 shows, to scale, the orbits of the planets Mercury, Venus, the Earth, Mars, Jupiter, Saturn, Uranus, Neptune and Pluto. The remaining bodies in Figure 1.3 are the large planetary satellites and the largest asteroid Ceres, which orbits between Mars and Jupiter. The upper scale in Figure 1.4, 150 million km, is very nearly the same as the *average* distance of the Earth from the Sun, which is always *very* close to 149.6 million kilometres (93.0 million miles). This distance used to define what is called the astronomical unit (AU), but because the value varies very slightly the AU is now nailed down as 149.5978715 million km (what precision!). It is an appropriate unit of distance within the Solar System. Note that 1 AU is nearly 24 000 times the radius of the Earth, or nearly 12 000 times its diameter, exemplifying how small the planets are compared with the distances that separate them.

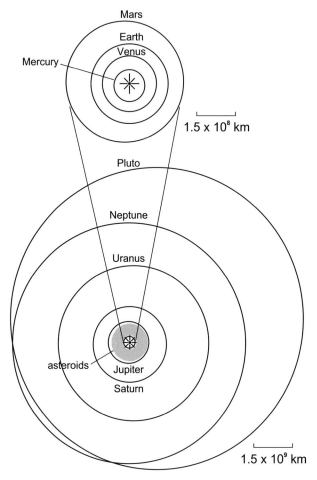

FIGURE 1.4 The Solar System; a face-on view of the planetary orbits. Though the orbits are not quite in the same plane, this makes no difference to the view on the scale here except for the orbit of Pluto, which would look slightly less circular in a face-on view (see Figure 1.5). The planets move around their orbits in an anticlockwise direction when viewed from above the Earth's North Pole.

The orbits of the planets are not quite circular; they are ellipses, which have the shape of a circle when it is viewed at an angle. The non-circular shape is not apparent on the scale of Figure 1.4. What *is* apparent, particularly for Mercury, Mars and Pluto, is that the Sun

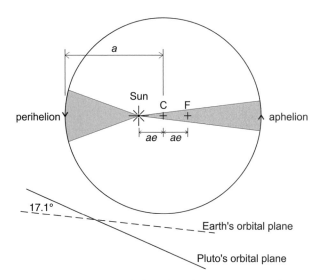

FIGURE 1.5 The orbit of Pluto, here face-on, to show what is meant by the semimajor axis, a, of an orbit and its eccentricity, e. C is the centre of the orbit. For Pluto $e = 0.251$. If e were zero the orbit would be a circle centred on the Sun which would be at C. The Sun is at one of the two points called the focuses of the ellipse. The other focus F is empty.

is not quite at the centre of the orbit. This is a consequence of their larger orbital ellipticity. To take a most pertinent example, Figure 1.5 shows the orbit of Pluto. Perihelion and aphelion are, respectively, the nearest and furthest points of the orbit from the Sun. Two quantities are shown that will be important in subsequent chapters. These are the size of the orbit as measured by its semimajor axis, a, and the non-circularity (ellipticity) of the orbit as measured by its eccentricity, e (a times e is shown in Figure 1.5). For Pluto $e = 0.251$, greater than for all the other planetary orbits. For a circle $e = 0$ and the Sun would lie exactly at the centre of such an orbit. (Note that the average distance of the Earth from the Sun that I referred to above, is the semimajor axis of the Earth's orbit.)

The orbits are also not quite in the same plane. Pluto's is the most inclined, at 17.1° with respect to the orbital plane of the Earth (Figure 1.5). The next most inclined planetary orbit is that of Mercury, at 7.0°. Inclinations are given the symbol i. The reference plane in the

Table 1.1 *The orbital elements a, e, i and P of the planets and the largest asteroid, Ceres (as of mid 2009).*

	Mercury	Venus	Earth	Mars	Ceres	Jupiter	Saturn	Uranus	Neptune	Pluto
a (AU)	0.387	0.723	1.000	1.524	2.766	5.203	9.515	19.24	30.20	39.64
e	0.206	0.0068	0.0167	0.0934	0.0793	0.0489	0.0534	0.0449	0.0095	0.251
i (°)	7.004	3.395	0.001	1.849	10.59	1.304	2.488	0.7719	1.769	17.14
P (years)	0.241	0.615	1.000	1.881	4.601	11.86	29.33	84.31	165.8	249.4

Data from the *Observer's Handbook 2009*. (The Royal Astronomical Society of Canada)

Solar System is called the ecliptic plane. At one time this was the orbital plane of the Earth, but as this plane tilts up and down *very* slightly with respect to the distant stars, the reference plane is now fixed in space.

From Figure 1.4 you might think that Pluto's orbit intersects that of Neptune, in which case they could collide! But the orbital inclination of Neptune is only 1.77°, so the orbits do not actually intersect. More on this in Chapter 2.

Table 1.1 gives the orbital elements a, e and i, and the orbital period, P, of each planet and of the largest asteroid (Ceres). There are slow, periodic variations in these elements, hence the 'as of mid 2009' in the table heading. They are caused mainly by the gravity of the bodies in the Solar System other than the Sun and the body in question. The excursions are small, except for the somewhat larger excursions of the values of e and i of Pluto. The slight variation in the Earth's inclination is apparent in Table 1.1: the value in mid 2009 was 0.001° rather than 0°.

The planets move around their orbits in the same direction, anticlockwise as viewed from above the Earth's North Pole; this is called the prograde direction. They move fastest near to perihelion because the gravitational pull of the Sun is greatest there, and they move slowest at aphelion, where the gravitational pull is least. More precisely, the line from the planet to the Sun sweeps out equal areas in equal time intervals, as illustrated by the two equal areas shaded in Figure 1.5. The time to go around an orbit once is called the orbital period, P. For the Earth it is one year (with respect to the distant stars), whereas for Pluto it is 249 years. Note that though the Sun pulls a body towards itself, the sideways motion of the body, dating back to the birth of the Solar System, turns what otherwise would have been an inward fall and an early demise, into orbital motion.

What is the relationship between the period, P, and the semi-major axis, a? As a increases, the distance around the orbit increases. For a circular orbit this distance is proportional to a, and so, if, for example, the value of a is doubled the distance around the circle is

BOX 1.2 KEPLER'S THIRD LAW OF PLANETARY
MOTION (FOR THOSE WISHING TO GO DEEPER)
In 1619 the German astronomer Johannes Kepler (1571–1630),
announced that P is proportional to $a^{3/2}$ i.e.

$$P = ka^{3/2}$$

where k is the constant of proportionality. This applies to circular and to elliptical orbits. That P increases as a increases is not
surprising – the orbit is bigger. However, this alone would make P
proportional to a. The extra sensitivity to a is because the speed of
the planet in its orbit decreases as a increases.

In the Solar System, if P is measured in years and a in AU
then the constant of proportionality has the value 1 exactly and
so $P = a^{3/2}$. For the Earth $a = 1$ AU and so the equation with $k =$
1 gives $P = 1$ year, which is correct! For Pluto, $a = 39.6$ AU and
so the equation gives $P = (39.6)^{3/2}$ years, which is 249 years, also
correct.

also doubled. If the speed of the planet in each orbit were the same,
then the period of the more distant planet would also be doubled.
However, because the force of the Sun's gravity decreases with distance, the speed in orbit also decreases, so that in doubling the value
of a, P more than doubles, in fact increasing by a factor of $2 \times \sqrt{2}$,
which is 2.828 . . . (to four figures). If a is increased three-fold then
P increases by $3 \times \sqrt{3}$, which is 5.196, and so on. Though I started
with circular orbits, these numerical results apply to elliptical orbits
too. The algebraic relationship between P and a constitutes Kepler's
third law of planetary motion, and for those of you wishing to go a bit
deeper please see Box 1.2.

Kepler's laws of planetary motion
We have now encountered three important laws of planetary motion.
These are called Kepler's laws after the German astronomer Johannes

Kepler (1571–1630), who announced the first two in 1609 and the third in 1619. They are based on accurate observations of the motions of the planets. The laws are as follows:

> *First law:* Each planet moves around the Sun in an ellipse, with the Sun at one of the two focuses of the ellipse (Figure 1.5).
>
> *Second law:* As the planet moves around its orbit, the straight line from the planet to the Sun sweeps out equal areas in equal intervals of time (Figure 1.5).
>
> *Third law:* P increases with a more rapidly than being proportional to a. (See Box 1.2 for the algebraic relationship.)

After their discovery, it was shown that these laws can be accounted for by fundamental physical theory, specifically Newton's laws of motion and law of gravity, developed in the second half of the seventeenth century by the English scientist Isaac Newton (1643–1727), but it would take us too far afield to go into details.

Small bodies in the Solar System: asteroids and comets

As well as planets, the Solar System also contains smaller bodies orbiting the Sun. More than a hundred thousand asteroids are known, mainly between the orbits of Mars and Jupiter. The asteroids are rocky or mixtures of rocky materials and iron, and up to several hundred kilometres in radius, but mostly much smaller. The largest asteroid is Ceres (Figure 1.3).

There are also two populations of bodies with a size range comparable to that of the asteroids but consisting of mixtures of rocky materials (including carbon-rich materials) and ices, mainly water ice. One population constitutes the Edgeworth-Kuiper belt, icy-rocky bodies in orbits of fairly low inclination, the semimajor axes of the great majority being between about 40 AU and 50 AU. The belt is also known as the Kuiper belt, and the members of the belt are known as Kuiper belt objects (KBOs). About 2000 Kuiper belt objects are known, with a steady flow of new discoveries. Chapter 6 is devoted to the Edgeworth-Kuiper belt.

The other population is the Oort cloud. The outer Oort cloud is a thick spherical shell of icy-rocky bodies, surrounding the Solar System, and extending from about 1000 AU (perhaps 10 000 AU) to the very edge of the Solar System at about 100 000 AU. The inner Oort cloud is more belt-like, and extends inwards from the outer Oort cloud towards the Edgeworth-Kuiper belt, perhaps coming as close as a few hundred AU from the Sun. Estimates of the total number of bodies in the Oort cloud range from about a million million (1 000 000 000 000), to ten million million (10 000 000 000 000).

Whereas the larger KBOs are close enough to be seen with telescopes, as tiny dots, Oort objects are much too far away. The existence of the Oort cloud is inferred from those of its members whose orbits are perturbed so that they travel through the inner Solar System, where they initially become fuzzy, and go on to develop huge, spectacular tails (Figure 1.6). They become comets.

Comets are divided into two main classes, the long-period comets, defined as having orbital periods greater than 200 years, and short-period comets which, it won't surprise you to learn, are defined as having orbital periods less than 200 years. As well as occupying different ranges of orbital period, the two classes differ in other ways. Long-period comets have orbital inclinations covering the full range, and thus bombard the Solar System from all directions. The short-period comets have no more than modest inclinations. In both classes, the fuzziness and the tails consist of gases evaporated from the icy component by the heat of the Sun, and of dust particles entrained in this gas flow.

Most of the short-period comets originate in the Edgeworth-Kuiper belt, from where they have had their orbits gravitationally perturbed so that they pass through the inner Solar System. A transitory population between the short-period comets and Kuiper belt objects constitute the Centaurs, which occupy unstable orbits with perihelia between Jupiter and Neptune. The remaining short-period comets, and all of the long-period comets come from the Oort cloud,

FIGURE 1.6 The comet Hale-Bopp, which visited the inner Solar System from the Oort cloud in 1997. (Francisco Diego, University College London)

from where they have been scattered into the inner Solar System by, for example, passing stars.

The eccentricities of the orbits of long-period comets are necessarily very large, with values approaching 1.000, a very narrow ellipse indeed, reaching from the inner Solar System out towards the Oort cloud. In some cases long-period comets depart on orbits that are not confined to the Solar System: they will escape into interstellar space and never return. The orbital eccentricities of the short-period comets are generally smaller, reflecting their closer origin.

(A third class of comet, the main belt comets, comprises a small number of bodies in low eccentricity orbits in the outer region of the asteroid belt. They develop weak dust tails near perihelion. They are thought to be icy asteroids that formed in the asteroid belt.)

Small bodies in the Solar System: natural satellites

Most planets have natural satellites. For example, the Earth has one, the Moon. Figure 1.3 shows the largest satellites. As well as the Moon, there are the four large satellites of Jupiter, in order of decreasing size, Ganymede, Callisto, Io (eye-oh) and Europa. Titan is the largest satellite of Saturn and Triton is Neptune's largest satellite, of which more in later chapters. Titania is the largest satellite of Uranus. You can see that two of the satellites are larger than Mercury and that six are larger than Pluto. These large satellites are not called planets because, in common with all satellites, they orbit a planet rather than occupying a separate orbit around the Sun.

1.3 PLANETARY COMPOSITIONS

External appearances of the planets

Figure 1.7 shows images of the planets, and the Moon, but omitting Pluto, presumably because no comparable image is yet available. This figure gives you a visual impression of appearance from a spacecraft. Comparison with Figure 1.3 shows that these images are *not* to scale.

Even the external appearance shows that these bodies differ from each other. A larger image would show that Mercury has an ancient surface, pock-marked with impact craters. It has only a tenuous atmosphere. Venus is shrouded in clouds in its thick carbon dioxide atmosphere, which overlies a hot surface displaying many volcanic features. The Earth, with its partly cloudy nitrogen-oxygen atmosphere, is the only body in the Solar System with large, exposed oceans of water. Our Moon, like Mercury, has an ancient, impact-scarred surface, but also less cratered regions, the Mare ('ma-ray'), which give rise to the 'The Man in the Moon'. Mars has a thin carbon

FIGURE I.7 Eight planets (not to scale). From top to bottom: Mercury, Venus, Earth, Mars, Jupiter, Saturn, Uranus, Neptune. The Moon is top right. (NASA/JPL-Caltech, PIA 03153) (See plate section for colour version.)

dioxide atmosphere covering a dry red-tinted surface. Larger images would show that roughly half of the surface is ancient cratered terrain, though streaked with water-carved channels that are evidence of warmer and wetter times in the distant past. Mars has polar caps of carbon dioxide ice and water ice.

Mercury, Venus, the Earth and Mars constitute the terrestrial planets. Beyond the asteroid belt we encounter the four giant planets, Jupiter, Saturn, Uranus and Neptune. These giants have very deep atmospheres, predominately hydrogen and helium, laden with clouds and hazes that reveal complex circulation patterns.

The interiors of planets

But what of the interior compositions and structures? Later in this chapter some methods of obtaining sizes and masses of planets will be outlined. Here I assume that we have these two values for a particular body. In this case its mean density is obtained by dividing its mass by its volume. This is an important constraint on its composition. Consider the density of some common materials. On the Earth's surface, under the pressure of the Earth's atmosphere, the density of pure liquid water is around 1000 kilogrammes per cubic metre (kg/m^3). Water ice is slightly less dense. Rocky materials are a few times more dense, around $3000 \, kg/m^3$ (with a few exceptions). Prominent among rocky materials are silicates, which are chemical compounds of silicon, oxygen and one or more metals (a chemical compound is a molecule consisting of two or more different chemical elements bound together). Oxides (oxygen-metal compounds) are also common. Pure iron has a density at the Earth's surface of $7873 \, kg/m^3$. Hydrogen and helium are gases at atmospheric pressure at the Earth's surface, but deep in the interiors of Jupiter and Saturn they can be compressed to liquids by the high pressures there.

My choice of these examples arises from the high abundance of their atoms in the Solar System (and indeed in the Galaxy and the Universe). Of the 92 naturally occurring chemical elements, hydrogen (H) is far and away the most abundant in the Solar System. Second comes helium (He), then the constituents of rocks, including oxygen (O), silicon (Si) and magnesium (Mg), and then iron (Fe). The water molecule, H_2O, consists of two abundant elements, two atoms of hydrogen, hence the subscript 2, and one of oxygen; water is an abundant substance. Iron always comes alloyed with about 6% nickel (Ni), called iron-nickel, or Fe-Ni. Its density (at the Earth's surface) is $7925 \, kg/m^3$. By rocky-iron I'll mean a mixture of rocky materials and

BOX 1.3 MEAN DENSITIES OF SPHERICAL BODIES
(FOR THOSE WISHING TO GO DEEPER)

The mean density d of a body is its mass divided by its volume. Consider a spherical body with a mass M and a radius R (the distance from its centre to its surface). The volume, V, of a sphere is $(4\pi/3)R^3$, where π is the Greek letter, pronounced 'pie'. In mathematics it is the circumference of a circle divided by its diameter, and has the value 3.14159... continuing indefinitely. Once we have V then we calculate d from $d = M/V$.

Suppose we have a plastic ball with a diameter of 0.200 metres (m). Therefore, $R = 0.100$ m, and $V = (4 \times 3.14159/3)(0.100$ m$)^3 = 0.00419$ cubic metres (m^3).

If its mass is 0.5 kg then its mean density is given by

$$d = 0.5 \text{ kg}/0.00419 \text{ m}^3 = 119 \text{ kg/m}^3.$$

The density of liquid water (at the Earth's surface) is about 1000 kg/m^3, so the ball could be solid and made of an extraordinarily low density plastic, or, more likely, it is hollow – remember that we have calculated the *mean* density.

iron-rich materials, be this Fe-Ni, iron oxides or iron sulphides (iron compounded with sulphur, another abundant element).

Turning to the planets, their mean densities are given in Table 1.2. The Earth comes top at 5520 kg/m^3. Mercury has the second highest mean density, 5430 kg/m^3. A rocky-iron composition is the only recipe that meets these mean density constraints. In fact, the proportion by mass of Fe-Ni is greater in Mercury than in the Earth, roughly 70% and 20% respectively. Mercury's slightly lower density is because the material in its interior is less compressed than deep in the Earth. This is because the greater the mass of a body the greater its internal pressures, and the greater the pressure the greater the density. Venus and Mars are also rocky-iron planets.

Without interior compression, hydrogen and helium would be present everywhere as gases, if they could be retained at all. The

Table 1.2 *The radii R, masses M and mean densities d of the planets and the largest asteroid Ceres.*

	Mercury	Venus	Earth	Mars	Ceres	Jupiter	Saturn	Uranus	Neptune	Pluto
$R(\text{km})^{[1]}$	2439.5	6052	6378	3396	478.5	71490	60270	25560	24765	1152
$M(M_E)^{[2]}$	0.05527	0.8150	1	0.1074	0.00021	317.8	95.16	14.54	17.15	0.00218
$d(\text{kg/m}^3)$	5430	5240	5520	3940	2700	1330	690	1270	1640	2040

[1] R is the equatorial radius. Planets are slightly flattened because of their rotation. The diameter is twice the radius.

[2] Masses, M, are given in units of the Earth's mass M_E, which is 5.9742×10^{24} kg. Except for Pluto, masses are known to somewhat greater precision than given here.

Data from the *Observer's Handbook 2009* (The Royal Astronomical Society of Canada), except for Ceres and Pluto. For Pluto see Chapter 3. For Ceres, see *Icarus* **72**, 507–518 (December 1987).

large early masses of Jupiter and Saturn have enabled these planets to acquire and retain hydrogen and helium and, moreover, in such quantities that these elements must constitute nearly all of the mass of these planets, as indicated by their low mean densities (Table 1.2). At the huge compressions in their interiors even water would give higher mean densities than the actual values of these planets. The mean density of Saturn is particularly low, because its mass is less than a third that of Jupiter, and consequently its interior is less compressed. Water, and rocky-iron materials, can contribute only a few percent to the total mass of Jupiter and Saturn.

Uranus and Neptune have masses between those of the terrestrial planets and Jupiter. Therefore, even though the mean densities of these three giant planets are not that different, the lower interior compressions in Uranus and Neptune indicate that hydrogen and helium are less dominant and that icy materials, notably water, predominate.

Finally, we come to Pluto. Its low density and low mass show that it is an icy-rocky body. This means that, as well as rocky-iron materials (which in Pluto's case might include carbon-rich materials), icy materials are present in abundance. In the case of Pluto, water and some other icy materials are thought to be present as solids, though others might be liquid. There is more on Pluto's composition and interior in Section 5.4 in particular.

Icy materials

In referring to icy materials I do not mean to imply that they are necessarily cold; 'icy materials' is the label for substances that, when solid, are icy in appearance and melt to become liquid, or sublime to become gas, at comparatively low temperatures. The most abundant icy material in the Solar System is water, H_2O, which melts at about 0°C (32°F). In fact, this temperature is slightly dependent on pressure. You might also have seen dry ice, which is solid carbon dioxide (CO_2), and at Earth's atmospheric pressure, at −78.5°C, does not melt but sublimes, i.e. turns directly from solid to gas, hence the name 'dry ice'. Only at pressures several times that at the Earth's surface does it melt. Other important icy materials in the Solar System include ammonia (NH_3), methane (CH_4) and carbon monoxide (CO).

Internal layering in planets

You might think that planetary interiors are uniform mixtures of their components. This is not the case. To detect internal layering the gravity of the planet has to be measured from close range at many points, ideally by an orbiting spacecraft. If such data are not available, clues can be obtained, for example, by detecting the strength of any magnetic field around the planet. A strong field is a powerful indication that liquid compounds rich in iron are present, and are probably concentrated towards the centre of the planet because of the high densities of such compounds. The details would take us too far afield. Suffice it to say that we do have pertinent data for all of the planets, except Pluto, which has not yet been sufficiently investigated. There will be more on Pluto's internal layering in Section 5.4. For the other eight planets here is a brief summary of what we know.

The terrestrial planets all show evidence that the intrinsically dense iron-rich component is concentrated towards the centre. As well as Fe-Ni, the core is likely to have small quantities of iron oxide or iron sulphide. This core is surrounded by a mantle of rocky materials, which is topped by a crust of other rocky materials of lower intrinsic density than those that comprise the mantle.

Jupiter is more homogeneous, to the extent that it might not have a core of intrinsically dense materials. The icy-rocky-iron materials seem to be dispersed fairly uniformly throughout the hydrogen and helium that dominate its composition. In contrast, the evidence shows that Saturn does have a core consisting of its intrinsically denser components. Overall, 5 to 10% of the mass of Jupiter consists of icy and rocky-iron materials. For Saturn the estimates are in the range 15–30%.

Uranus and Neptune certainly have cores beneath their massive atmospheres of hydrogen and helium. Estimates of the percentage of the planetary mass in these atmospheres range from 5 to 15%. The cores consist of icy and rocky-iron materials.

Heat sources and interior temperatures in planets

Planetary interiors range from cool, e.g. somewhere between about −150°C and about 300°C at the centre of Pluto, to hot, e.g. 5200°C at the centre of the Earth, to very hot, e.g. somewhere in the range 15 000–21 000°C at the centre of Jupiter. Temperatures decrease with decreasing depth, but only gradually. In the four giant planets the internal temperatures ensure that the icy and rocky materials are liquid. Pluto's interior is cool, so icy materials, or some at least, would be solid there, except perhaps very near the centre – more on this in Section 5.4.

What are the heat sources that have created and maintain the internal temperatures? The main ones are:

- heat from the kinetic energy (energy of motion) in the materials that came together to form the planet
- heat released by the decay of radioactive isotopes (some disappear soon after planetary formation, others are still active)
- heat released if the denser materials, because of the gravitational attraction between them, settle towards the centre of the planet, in a process called differentiation; as the materials settle, the downward motion is dissipated in the form of heat
- heat from changes in the tidal distortion of the planet caused by the gravity of a nearby body, e.g. the (small amount of) heating of the Earth's interior by the Moon when its tidal distortion of the whole Earth (not just the oceans) sweeps through the Earth as it rotates
- heat released if a liquid component solidifies (this is called latent heat).

The temperatures in the interior are determined by the balance between heat gains from heat sources, and heat losses. A homely analogy is filling a bath with water. The depth of the water is the analogue of temperature. Heat sources are represented by the tap filling the bath. Let's suppose that the person filling the bath has been care-less and has not put the plug in the drain hole. If the hole is small the water will begin to fill the bath, but the rate of escape increases as the water deepens. It is possible that an equilibrium will be reached, in which the rate of escape equals the rate of input of water from the tap. The water level is then the analogue of the equilibrium temperature.

To return from the bathroom, the heat losses from a planet are determined by the rate at which heat is conducted to the surface, and the rate at which heat escapes from the surface to space. This is a complicated subject but there is one fairly general rule: the reservoir of heat sources is larger the more massive the planet, and the rate of heat loss is greater the larger its surface area. The more massive planets usually have the larger volumes, so, crudely speaking, the heat reservoirs increase as R^3, where R is the radius of the planet. Equally crudely, the rate of heat loss increases as R^2. This means that, all else being equal the cooling rate increases as R decreases – small bodies cool faster than large bodies. To take an extreme example, it is expected that Pluto has cooled far more rapidly than Jupiter. I emphasize that this is a very rough rule of thumb, because all else is *not* equal, particularly because the rate of transfer from the interior to the surface differs considerably from one planet to another. (If you want to learn more about this, and other topics, you should scrutinise the books on the Solar System in *Further reading and other resources*.)

The internal temperatures determine the level of activity in the deep atmospheres of the four giant planets, and the level of geolog-ical activity in the terrestrial planets: the higher the temperatures the greater the activity in both cases. The internal temperatures also determine whether a planet has a strong magnetic field, a requirement for this being a liquid electric conductor, such as iron-rich compounds, which require high temperatures.

When all internal heat sources have long been exhausted, the whole body will be at a uniform temperature throughout, determined by the radiation it absorbs from the Sun, and by how much of the heat from this absorbed radiation is radiated back into space. However, none of the planets in the Solar System, not even tiny Pluto, has reached this state yet.

The Solar System, of course, does not contain only planets and a host of smaller bodies, but also the Sun, which is far and away the most massive body in the Solar System, keeping its retinue in orbit around it, and warming planetary surfaces.

1.4 THE SUN

Without the Sun, the surface of the Earth, and of the other three terrestrial planets (Mercury, Venus and Mars), would be at temperatures far below 0°C. Even though the internal heat sources of the Earth could keep the deep interior warm, the upper few hundred metres and the surface, except in volcanically active regions, would be far too cold for life, and the oceans would almost everywhere be frozen. Life at and near the surface would be impossible over nearly all of the Earth.

The surface of the Sun, called its photosphere, has a temperature of about 5500°C. This is called the Sun's effective temperature. It is this high surface temperature of a body with such a large surface area that makes the Sun luminous enough to warm the surfaces of the terrestrial planets and the upper atmospheres of the giant planets, and even the icy surface of distant Pluto. It does so through the electromagnetic radiation it emits from its photosphere, particularly at visible and infrared wavelengths. This has been the case throughout the 4600 million years since the Sun and the planets were born. What could have made the Sun so luminous over such an enormous span of time? The answer lies deep in its interior.

The solar interior

The Sun has a radius of 696 265 km, which is about 100 times that of the Earth. Therefore, the Sun has a volume about a million times greater than the Earth's. The Sun is a very dense gas in its interior. It consists almost entirely of the lightest two chemical elements, hydrogen and helium. When the Sun was born, these elements accounted

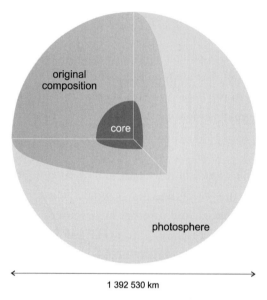

FIGURE 1.8 The Sun, showing the core where hydrogen fusion occurs.

respectively for about 71% and 27 to 28% of the Sun's mass, leaving rather less than 2% for all the other 90 chemical elements. By contrast, the terrestrial planets, being rocky-iron, consist almost entirely of these other elements.

Figure 1.8 shows a cut-away of the Sun. Temperatures increase rapidly with depth, and are everywhere high enough for the atoms to have lost all of their electrons to become ions, in a process called ionisation. Within the core the temperatures are high enough for a process called thermonuclear fusion to occur, indeed this is what defines the core. Nuclear fusion is the process in which two atomic nuclei are hurled together so fast that the nuclei join together to make a different nucleus. The higher the temperature the greater the speed of the nuclei, and the more often nuclear fusion occurs, hence the prefix 'thermo'. In the Sun's core, the overall effect of a series of thermonuclear fusion reactions is to convert four hydrogen nuclei into one helium nucleus – four protons have been joined to make a nucleus consisting of two protons and two neutrons. In this process a tiny amount of energy is released, in the form of gamma rays, fast

BOX 1.4 THE STRUCTURE OF THE ATOM AND THE
CHEMICAL ELEMENTS (PLEASE READ)

The atom consists of a tiny nucleus made up of a certain number of electrically neutral particles called neutrons, and a certain number of positively charged particles called protons. The nucleus is surrounded by a swarm of tiny negatively charged particles called electrons. For an atom to be electrically neutral the number of protons has to equal the number of electrons, otherwise it is called an ion. A proton and a neutron are each around 1837 times the mass of an electron, so nearly all the mass of the atom is in the nucleus. By contrast, the volume of the atom is determined by the electrons, which swarm in a volume far larger than that of the nucleus. Atomic radii are so small that, placed side by side, several million atoms would be needed to span a millimetre.

The nucleus is typically about 10000 times smaller than the orbits of the outer electrons. To put this into a human scale, imagine that a nucleus had a diameter of 0.1 millimetres – a tiny pinhead. The electron orbits would then extend out to about 1 metre.

The simplest atom is that of hydrogen, denoted by H. It consists of a nucleus of one proton surrounded by one electron. This makes it the least massive atom and the least massive nucleus of all. Next is deuterium, in which the nucleus consists of one proton and one neutron, so it is about twice the mass of hydrogen. Like hydrogen, the deuterium nucleus is surrounded by one electron. The chemical properties of an atom are determined by the number of electrons in its neutral form, which is equal to the number of protons in the nucleus. The number of protons, called the atomic number, defines a chemical element. Thus, hydrogen and deuterium are the same chemical element, which is called hydrogen. Hydrogen and deuterium are called different *isotopes* of hydrogen.

The next element must have two protons in its nucleus. This is helium (He). The common isotope has two neutrons in the nucleus, making it four times as massive as H.

And thus we build up the chemical elements, with no gaps in the number of protons. As a final example consider carbon (C), which has six protons in its nucleus. The most common isotope by far has six neutrons, making 12 nuclear particles in all. This is called carbon-12, or ^{12}C. There are also isotopes with 4, 5, 7, 8, or 9 neutrons: ^{10}C, ^{11}C, ^{13}C, ^{14}C and ^{15}C.

Not all isotopes are stable. Some decay into other isotopes. Such unstable isotopes are said to be radioactive; they emit particles from the nucleus, for example neutrons, or gamma rays. Among the carbon isotopes only ^{12}C and ^{13}C are stable.

positrons, and the almost mass-less electron-neutrinos. (A positron has the same tiny mass as an electron, but is positively charged.) The electron-neutrinos are very unreactive, and almost all of them escape from the Sun without depositing any energy. It is the positrons, electrons, and particularly the gamma rays, that carry heat outwards.

At temperatures above about 10 million degrees Celsius the fusion of hydrogen to helium in the Sun's dense core happens so often that a lot of energy is released. The core of the Sun has always been hotter than this (it is currently about 14 million degrees Celsius ($^{\circ}$C)). Consequently, since its birth, the helium proportion in the Sun's core has increased at the expense of hydrogen.

It is the core fusion that makes the Sun so luminous. The heat released in the core gradually diffuses outwards, and is radiated into space from the photosphere. This electromagnetic radiation has warmed the Earth's surface throughout its history, and has been essential for the presence of surface life. The Sun's luminosity is not constant. On a timescale of roughly a million years (Myr) the variations are slight, but since the Sun was born 4600 Myr ago its luminosity has increased from about 70% of its present value. Its luminosity

BOX 1.5 A COSMIC CLOCK-TICK (PLEASE READ)
A million years is written 1 Myr, where 'M' stands for 'million' and 'yr' stands for 'year'. It is a long time compared with a human life-span, but it is a convenient unit of time when we are considering the history of a planet or a star. You can think of 1 Myr as the tick of a cosmic clock, and so there have been 4600 clock ticks since the Earth was born. The Myr unit of time will be used throughout this book.

will continue to increase until in a few thousand Myr from now the Earth will have become too hot for life. Further in the future, about 6000 Myr from now, the hydrogen in the core will all have been fused to helium; this marks the end of what is called the main sequence lifetime of the Sun, which began at its birth. The Sun will then start on its transition to becoming a red giant star, and only the outermost Solar System will be habitable.

1.5 THE ORIGIN OF THE SOLAR SYSTEM

The Solar System formed 4600 Myr ago. This age has been determined by the radiometric dating of components in primitive meteorites, small bodies that impact the Earth but survive to be collected. (For further details see the books on the Solar System in *Further reading and other resources*.)

In recent decades, because of its explanatory power, the so-called nebular theory has been widely adopted to explain the origin of the Solar System. In this theory the Solar System formed from a cloud of gas and dust. The gas consisted mainly of hydrogen and helium, and the dust mainly of the other 90 chemical elements. As the cloud contracted, its rotation rate increased and it flattened to form a thick disc (Figure 1.9). The dust settled through the gas to the mid plane of this disc, to form a thin dust disc. In its inner region, near where the Sun was forming at the centre of the disc, the dust disc consisted mainly of iron-rich compounds, and other compounds with high melting points, notably silicates. Further out, where temperatures were

FIGURE 1.9 An artist's impression of the disk of gas and dust from which the Sun and the planets formed. The Sun is forming at the centre. This is an oblique view – the disc is circular. (Julian Baum, Take 27 Ltd)

lower, carbon-rich compounds were also present in solid form. Even further out, beyond what is called the ice line, as its name suggests, water ice was present in the dust. Well beyond the ice line other icy materials were present in the dust.

The Earth and the other terrestrial planets formed from the inner region of the dust disc, which is why they are rocky-iron in composition. The first step in planet formation was the agglomeration of some of the dust to form millimetre sized objects, which gradually accreted material to form kilometre sized bodies called planetesimals. These collided, causing disruption but also growth through further accretion. The outcome was a set of what are called planetary embryos, ranging in diameter from a few hundred to a few thousand kilometres. In the subsequent 100 Myr or so, by far the slowest stage in the process, the set of embryos interacted with each other until just a few large bodies were left – the terrestrial planets. The energy liberated in their formation caused much of each planet to melt, to form an Fe-Ni core, perhaps with some oxygen or sulphur, surrounded by a silicate mantle.

There were also small quantities of other substances. In the case of Venus, the Earth and Mars these subsequently gave the planets their early atmospheres, and in the case of the Earth, its oceans. A proportion of these substances came with the dust and larger bodies throughout the formation of the terrestrial bodies, and a proportion came near the end of their formation, from bodies rich in water and carbon compounds, analogous to present day comets and certain asteroids. These late arrivals contributed to a heavy bombardment that in the case of the Earth continued to about 3900 Myr ago, i.e. to about 700 Myr after the Earth had acquired nearly all of its mass.

In the outer Solar System things were different because beyond the ice line water ice was present, the most abundant icy material in the Solar System. This, along with rocky materials, allowed kernels to form with several times the mass of the Earth, massive enough to capture hydrogen and helium gas from the disc. This gas contracted to form massive hot envelopes surrounding the hot cores, all much quicker than the formation of the terrestrial planets. This is the favoured model for the formation of the giant planets.

Capture of gas ceased when the disc around the Sun was dissipated. This is thought to have been due to a burst of intense solar activity called the T Tauri phase, in which a copious fast wind of particles, mainly protons and electrons, was emitted by the Sun. Uranus and Neptune formed more slowly than Jupiter and Saturn because of their greater distance from the Sun and the consequent lower density of the disc. However, at their present distances from the Sun their cores would only have formed after the disc was dissipated, which would have left them bereft of the thick atmospheres rich in hydrogen and helium that we know they possess. Fortunately there are computer simulations that show that Uranus and Neptune could have formed closer to Saturn than they are today, and thus would have been able to capture gas, though less than Jupiter and Saturn, as observed. They then migrated outwards, through a variety of subtle gravitational interactions with the remnants of the disc and with remnant planetesimals. Planetary migration was also important in sculpting the Solar System beyond Neptune, as you will see in later chapters.

A planet failed to form in the space between Mars and Jupiter very probably because of the gravitational disturbance caused by Jupiter that stalled the formation of bodies much larger than a few hundred kilometres across. The asteroids are thus much smaller than the planets. The present population of asteroids is different from the original population, which has been modified by collisions between them. This has changed the size distribution. Mass has also been lost, not only through collisions but also through the continuing gravitational influence of Jupiter, and to a lesser extent of Mars.

Pluto remained small probably because of the paucity of solid materials in its distant feeding zone, and increased collision speeds caused by the gravity of the giant planets. The materials were icy and rocky, so Pluto stalled as a small icy-rocky body.

Satellites, rings, the Oort cloud and the Edgeworth-Kuiper belt
Planetary satellites have been formed in a variety of ways. The Earth's Moon is probably the result of the impact of a Mars sized embryo late in the Earth's formation. Some of the mass of this body would have joined the Earth, the rest, ejected from the Earth's mantle, would have formed a disc of debris around the Earth, from some of which the Moon coalesced. Mars's two tiny satellites are probably captured asteroids. The inner satellites of Jupiter, Saturn and Uranus, which include all but one of the big satellites, are thought to have formed from discs of dust and rubble around the planets, somewhat like the formation of the terrestrial planets around the Sun. Many of the other satellites of the giant planets, particularly the small ones far from their planet, are thought to have been gravitationally captured. This is also the case for Neptune's large satellite, Triton.

All four giant planets are encircled by rings, that of Saturn being by far the most extensive at present (Figure 1.7, though the illustration does not show a tenuous ring recently discovered much further out). The rings consist of small solid particles in thin sheets fairly close to the planets and in their equatorial planes (except for the recently discovered ring of Saturn). In the main, the particles are tiny fragments of larger bodies that came sufficiently close to the giant planet to be disrupted by its powerful gravity. Particles are gradually lost from the rings, but a large supply of new particles would arise from the disruption of a small satellite that strays too close to the giant. Presumably this happened relatively recently in Saturn's vicinity, which is why it presently has the most extensive ring system.

After the formation of the planets, there was plenty of material left over, from dust, to small rubble, to planetesimals. In the inner Solar System the composition would have been predominantly rocky-iron. In the outer Solar System icy materials would also be present. The Oort cloud is thought to consist of objects that were flung outwards by the gravity of Uranus and Neptune, sometimes with the aid of Jupiter's and Saturn's gravity. Many such objects would have been sent into interstellar space, but others, the Oort cloud objects, narrowly escaped this fate.

The members of the Edgeworth-Kuiper belt are thought to have formed beyond Neptune, the orbits of the ones nearer to Neptune having been greatly modified by Neptune's gravity, particularly during its outward migration. Some of the more distant Kuiper belt objects are also thought to have been emplaced by Neptune's gravity, which has also been responsible for stirring their orbits to an extent that has contributed to the prevention of growth much beyond Pluto's size. Pluto is now regarded as a comparatively large Kuiper belt object. Chapter 6 is devoted to the Edgeworth-Kuiper belt.

That concludes my description of the Solar System and how it formed. In this chapter it just remains for me to outline how distances, sizes and masses are obtained for Solar System bodies.

1.6 MAKING MEASUREMENTS OF DISTANCES, SIZES AND MASSES

This section contains boxed material for readers comfortable with basic algebra. The unboxed material takes a much more qualitative approach, but will equip you with the basic ideas.

Distances

Kepler's third law (Section 1.2) enables us to obtain *relative* distances in the Solar System, because it relates the orbital period of a body to its orbital semimajor axis. Thus, if we measure the orbital periods of bodies A and B, then we can obtain the ratio a_A/a_B of the semimajor axes of their orbits. If one of the two bodies is the Earth, for which the semimajor axis is 1 AU, then we can express the other semimajor axis in AU. This can be repeated for all orbits. Moreover, from the shape and orientation of the orbits, we can draw a scale plan of the Solar System, and from our knowledge of how bodies move around their orbits we can show where the various planets and other bodies lie at each instant. At this instant we can thus express in AU the distance between any two bodies in orbit around the Sun.

But how do we establish the value of the AU in everyday units, such as the kilometre or the mile?

For many years the distance from the Earth to another body has been obtained by triangulation, as illustrated in Figure 1.10. The direction to a particular point on the body is measured from two points on the Earth of known separation (obtained from surveying methods). The difference between the two angles is called the parallax of the body. By drawing the thin triangle in Figure 1.10 to scale the distance to the body can be found. For example, if the baseline, B, in Figure 1.10 is the length of the Earth's diameter, 12 756 km, then the parallax of some point on the Sun when the Earth is at perihelion is 0.00487°. This is a very small angle; the angular diameter of the Moon in our skies is around 0.5°. Small angles are more conveniently expressed in seconds of arc. There are 3600 seconds of arc in a degree of arc and

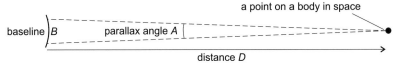

distance = actual size /angular size

FIGURE 1.10 Obtaining the distance, D, to a body by a measurement of the angle, A, from the baseline length, B, perpendicular to the line of sight. The formula in the figure requires that A be measured in radians (a radian is $57.3°$ to three significant figures).

therefore the Sun's parallax from a base equal to the Earth's diameter is 17.5 seconds of arc, or 17.5 arcsec.

Trigonometry is used rather than scale drawings, for all sizes of angles, where simple formulae relate the various angles and distances.

Though, in principle, the distance from the Earth to the Sun, and therefore the astronomical unit, could be determined in the manner just outlined, more accurate values can be obtained using smaller bodies. This is because identifying the same point on the Sun as seen from two well-separated observatories cannot be done as accurately as identifying a particular small body.

A suitable small body was discovered in 1898, an asteroid subsequently named Eros. This tiny body (now known to be non-spherical, $34.4 \times 11.2 \times 11.2$ km) approached the Earth in 1975 as close as 0.15 AU, and it often comes almost as close. Clearly, the closer the body the greater its parallax and the more accurately its distance from the Earth in kilometres can be determined. With accurate knowledge of its orbit, its distance from the Earth in AU can be obtained for any time and thus an accurate value for the number of kilometres per AU can be derived. Values were obtained in this way in 1931, 1950 and 1968. The 1968 value, from J H Lieske at the California Institute of Technology, is 149.60040 million kilometres.

Today, the astronomical unit is best measured using radar reflections. Radar pulses travel at the speed of light, c, which is known very accurately. Time intervals can also be measured very accurately, so if we measure the time interval, t, between sending a radar pulse

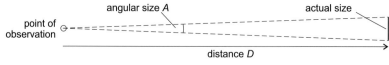

actual size = angular size × distance

FIGURE 1.11 Obtaining the actual size of an object from its angular size, A, and its distance, D. The formula in the figure requires that A be measured in radians (a radian is 57.3° to three significant figures).

from the Earth to a planet and receiving its reflection, then the distance from the Earth to the planet is $ct/2$. Spacecraft receiving and returning radio messages have also been used. The current value for the AU is 149.5978715 million kilometres.

Sizes

I'll restrict the discussion to bodies that have discs discernable in a telescope; Section 3.1 outlines how sizes are obtained when no disc is discernable.

Figure 1.11 shows that to obtain the actual size (for example in kilometres) we need to know the angular size of the body and its distance. You have seen in the preceding subsection how the distance is obtained, so in principle the actual size can be obtained from a scale drawing. As with distances, trigonometry is used rather than scale drawings, for all sizes of angles, where simple formulae relate the various quantities.

Masses

To determine the mass of a body we need to measure its gravitational effect on another body. I'll restrict myself to cases where the other body has a much smaller mass than the body of interest. An example is a spacecraft passing near a planet. The gravitational attraction of the planet will bend the spacecraft's orbit towards the planet, and from the change in the spacecraft's path the mass of the planet can be calculated. Another example is the orbit of a small natural body around another, such as the orbit of a planet around the Sun, from

BOX 1.6 DISTANCES, SIZES AND MASSES (FOR THOSE COMFORTABLE WITH ALGEBRA)

Distances

Kepler's third law enables us to obtain *relative* distances in the Solar System. Thus, if we measure the orbital periods, P, of bodies A and B, then the ratio of the semimajor axes a of their orbits is obtained from the equation in Box 1.2, $P = ka^{3/2}$:

$$\frac{a_A}{a_B} = \left(\frac{P_A}{P_B}\right)^{2/3}.$$

If one of the two bodies is the Earth, then we can express the other semimajor axis in AU. This can be repeated for all orbits.

To obtain a distance in AU in kilometres, we seek a formula relating D to A and B in Figure 1.10. Given that D is very much greater than B a simple formula can be used:

$$D = B/A.$$

This formula requires that A be measured in radians. This is a unit such that there are 2π radians in $360°$, where π is the circumference of a circle divided by its diameter, and has the value $3.14159\ldots$ This means that one radian is $360°/2\pi$, i.e. $57.2957\ldots°$.

Sizes

In this case we know D and A and wish to calculate B. A rearrangement of the simple formula above is what is needed

$$B = D \times A,$$

where again A is measured in radians.

Masses

Consider a planet with a mass M orbited by a single satellite with a mass m, then from Newton's laws of motion and gravity we get

$$M + m = \frac{4\pi^2 a^3}{GP^2},$$

where a is the semimajor axis of the orbit of m with respect to M, P is the orbital period of each around the other, and G is the gravitational constant, which is the same regardless of the masses. If m is much less than M, it can be omitted from the equation and we get the value of M without knowing the value of m. If m is *not* much less than M then their motions around the centre of mass of the two bodies has to be determined, in which case values for both masses are obtained – see Box 3.2.

which the mass of the Sun can be obtained, or the orbit of a small satellite around a planet, such as Europa around Jupiter, from which the mass of Jupiter can be obtained.

That concludes Chapter 1. As I said at the beginning, much of what you have learned will support and enhance your understanding of subsequent chapters. The next chapter tells the story of the discovery of Uranus, Neptune and then Pluto.

2 The discovery of Uranus, Neptune and Pluto

Before I tell you the story of Pluto's discovery, it is both instructive and relevant to the discovery of Pluto for you to learn, briefly, about the discovery of Uranus and, in more detail, about the discovery of Neptune. Between them, these three planets were found by strikingly different means.

2.1 THE DISCOVERY OF URANUS

Until 1781 just five planets were known: Mercury, Venus, the Earth, Mars, Jupiter and Saturn, all readily visible to the unaided eye. Then, on 26 April of that year a scientific paper was read to the Royal Society that opens as follows.

> 'On Tuesday the 13th of March, between ten and eleven in the evening, while I was examining the small stars in the neighbour-hood of H Geminorum, I perceived one that appeared visibly larger than the rest: being struck with its uncommon magnitude, I compared it to H Geminorum and the small star in the quartile between Auriga and Gemini, and finding it so much larger than either of them, suspected it to be a comet.'

This is the opening of a paper written by the Germano-British astronomer William Herschel (1738–1822). It was read to the Royal Society by the British physician William Watson (1744–1825). Thus was announced to the world the discovery, not of a comet, but of what was soon shown to be a planet. Its true nature was clear by May 1781 after its large, low eccentricity orbit had been established. The name 'Uranus' was first suggested by the German astronomer Johann Elert Bode (1747–1826) in 1784, but this name was not everywhere accepted

FIGURE 2.1 Left: William Herschel in 1814. (Institute of Astronomy, Cambridge UK) Right: a replica of the telescope with which Herschel discovered Uranus. (Herschel Museum of Astronomy, Bath, UK, by permission) (See plate section for colour version.)

for several decades. In mythology Uranus is the father of Saturn and the first ruler of Mount Olympus.

Uranus is the first planet to have been discovered in recorded history, and although it had been recorded as a star on several earlier occasions, it is not hard to understand why it remained unrecognised for so long. It is only 42% of Saturn's diameter and is twice as far away. Consequently it is only just about visible to the best unaided eyes in dark, clear skies. Moreover, the large semimajor axis of its orbit means that it takes a little over 84 years to complete an orbit

FIGURE 2.1 (cont.)

and therefore moves extremely slowly against the stellar background. Consequently, it is easily mistaken for a faint star.

But Uranus is readily detected in a small telescope, and under modest magnification it can be seen as a disc, unlike the stars, which are so far away that they look like points of light even at high magnification. Herschel discovered Uranus in Bath, England, in the back garden of his house at 19 King Street, using a reflecting telescope that he had made himself (Figure 2.1). This has a main mirror just 6.2 inches (158 mm) diameter, a size comparable to the reflecting telescopes used by many amateur astronomers today. At the time, he was making a systematic survey of the stars with particular emphasis on cataloguing double stars, that is, two stars separated by a very small angle on the sky (either because they are actually close together in orbit around each other, or because they are nearly in the same line of sight, the one lying far beyond the other).

He discovered Uranus with an eyepiece that gave an angular magnification of 227, i.e. increased the angular diameter 227 times. He changed eyepieces to increase the magnification to 460, and then to 932. The disc of Uranus increased in size accordingly, whereas the stellar images were not enlarged, their apparent angular diameters being determined by the resolution limit of Herschel's telescope.

William Herschel continued his astronomical observations, mainly of stars and galaxies, but in 1787, with a 48-inch (1.2-metre) diameter reflecting telescope in Slough, England, he discovered two satellites of Uranus, subsequently named Titania and Oberon by William's son, John (1792–1871), who was also an accomplished astronomer. The orbits of the satellites enabled the mass of Uranus to be calculated (Section 1.6). Though Uranus is a substantial body, 14.54 Earth masses (M_E), its mass was too low for it to have caused observable discrepant motions of Jupiter and Saturn, which would have led to the prediction of a planet beyond Saturn.

With the unanticipated discovery of Uranus, the question arose 'Are there planets beyond Uranus?'.

BOX 2.1 TELESCOPES (PLEASE READ)

There are two basic sorts of telescopes: refractors and reflectors. The main imaging element of the former is a large lens, whereas the latter has a large concave mirror (Figure Box 2.1). Binoculars are essentially two refractors mounted side by side.

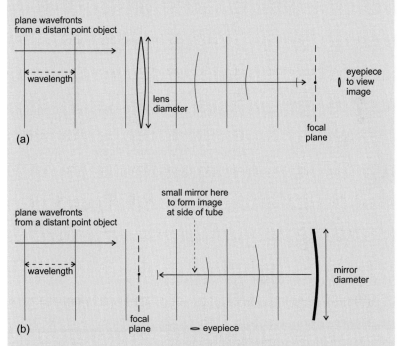

(a)

(b)

FIGURE BOX 2.1 Simplified drawings of (a) a refractor (b) a reflector. In the latter case the image plane needs to be made accessible, for example by means of a small plane mirror to deflect the light to the side of the tube, in a version called a Newtonian reflector.

The crucial optical feature is the size of the main lens or mirror, the larger the better, but also the more expensive and the more difficult to make with sufficient accuracy. The area of the main lens or mirror that gathers light is called the aperture. The larger the aperture the greater the amount of light gathered from the object under scrutiny. This enables fainter objects to be seen. Also, the larger the aperture the greater the amount of detail visible, i.e. the better the resolution. This is because the image formed of a distant point of light, even with perfect optics, is not a point, but a fuzzy disc; the larger the main lens or mirror, the smaller the disc. This is called the diffraction limit (to resolution). But atmospheric turbulence can make things worse. Even in the steadiest, clearest skies it is the atmosphere that determines the resolution for telescopes with aperture diameters exceeding about 300 mm, though the increased light-gathering power remains a big advantage.

The magnification of a telescope with a given main lens or mirror is determined by the eyepiece. The purpose of the eyepiece is to magnify the image produced in the focal plane of the main element (Figure Box 2.1) to enable the human eye to perceive the detail. The greater the magnification the more detail becomes visible, up to the diffraction limit or the limit imposed by the atmosphere. Further magnification enlarges the image, but no further detail is obtained and the field of view and the image brightness decrease.

2.2 THE DISCOVERY OF NEPTUNE

In the midst of the French Revolution in the turbulent Paris of 1789 the French astronomer Jean Baptiste Joseph Delambre (1749–1822) was preoccupied with the orbit of Uranus.

Delambre had available numerous observations of Uranus spanning the period from its discovery in 1781 up to 1788, and he had also pre-discovery observations of Uranus that by then had been unearthed. The earliest were six observations made in 1690 by the British astronomer John Flamsteed (1646–1719). Another was made in 1756 by the German astronomer Tobias Mayer (1723–1762). Neither Flamsteed nor Mayer had suspected that that particular point of light was a new planet.

The pre-discovery observations were of particular importance because they greatly extended the period over which the positions of Uranus were known. Consequently its orbit could be calculated more accurately than with the post-discovery observations alone, as could the position of the body in its orbit at any given time. Delambre used the observations, along with Newton's Laws of Motion and Law of Gravity, to calculate the orbit of Uranus, making due allowance for the perturbing effects of the gravity of Jupiter and Saturn using their orbits and masses (the latter determined from their satellites' orbits). The positions that Delambre calculated matched the observed positions of Uranus to within the observational uncertainties. This meant that a table of the future positions of Uranus could be calculated. Delambre published such a table in 1790.

The happy agreement between's Delambre's positions for Uranus and the observed positions lasted until 1820. It would have lasted *less* long had not Europe been in political turmoil for much of the 30 years after 1790, and had not the astronomical community largely lost interest in Uranus in that period. This loss of interest stemmed partly from the passing of novelty, and partly from the growth of interest in the apparently empty gap between Mars and Jupiter, where the empirical Titius-Bode rule predicted that there should be a planet.

The Titius-Bode rule

If we compare the semimajor axes of the planets known in 1790 then something curious emerges. One way of making this comparison is shown in Figure 2.2. The planets have been numbered in order from the Sun: Mercury is numbered 1, Venus 2, Earth 3, Mars 4, Jupiter 6, Saturn 7 and Uranus 8. The semimajor axes, a, of the orbits have been plotted versus each planet's number, on what is called a logarithmic scale, in which equal distances along the scale correspond to equal multiples, rather than to equal differences as on the more familiar scales. The curious thing is that the data lie close to a straight line, as shown. This means that the semimajor axes increase by about the same multiple each time we go from one planet to the next one out. This is one of several ways of expressing the Titius-Bode rule, named after the German astronomers Johann Daniel Titius (1729–1796) who

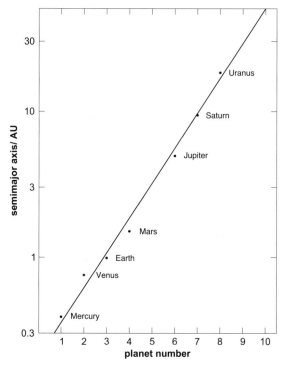

FIGURE 2.2 The semimajor axes, *a*, of the planets known up to 1801, versus their order *n* from the Sun: $n = 1$ Mercury, 2 Venus, 3 Earth, 4 Mars, 6 Jupiter, 7 Saturn, 8 Uranus. (The rule is usually expressed as an equation relating *a* to *n*, of which there are several versions.)

formulated a version of it in 1766, and Bode who published it in 1772, a few years before the discovery of Uranus, though the rule played no part in the discovery.

You can see that the planets only lie close to the line because number 5, at about 2.8 AU, is left unoccupied. This gap was filled on New Year's Day 1801 by the discovery of the asteroid Ceres ('series') in an orbit with a semimajor axis, *a*, close to 2.8 AU. Ceres is the largest asteroid (478.5 km radius), now known to be in an orbit with $a = 2.766$ AU. By 1807 three more had been discovered, Pallas (2.773 AU), Vesta (2.362 AU) and Juno (2.672 AU). Pallas is the second largest asteroid. Vesta comes third and Juno ninth (137 km radius). It was disappointing that a decent-sized planet had not been found, but at least the four asteroids then known were more or less at the distance from the Sun predicted by the Titius-Bode rule. We know now that few of the hundreds of thousands of known asteroids are concentrated around 2.8 AU.

Today the rule is regarded as no more than a qualitative statement that the spacing between orbits tends to increase with increasing distance from the Sun. This is a feature of nebular theories of Solar System origin (Section 1.5). But in the nineteenth century it played a role in the discovery of Neptune.

A problem with Uranus

To pick up the story of Uranus again, by 1820 there were observations of the planet spanning nearly 40 years since its discovery, and in addition the number of pre-discovery observations had grown to seventeen. In that year the French astronomer Alexis Bouvard (1767–1843) attempted to correct his tables of the positions of Jupiter, Saturn and Uranus.

Bouvard was able to obtain orbits for Jupiter and Saturn that fitted all the observations. But he was *unable* to do the same for Uranus. He was also aware that the position of Uranus predicted by Delambre differed markedly from the more recent observations. One of Bouvard's orbits fitted the older observations; a different orbit fitted the more recent ones. Bouvard resorted to rejecting the older observations as inaccurate, even though this meant attributing errors of about 50 arcsec to astronomers who were generally regarded as having attained observational accuracies of about 5 arcsec. Bouvard published his new tables for Uranus in 1821, but even though he had rejected the older observations he was only able to match the more recent observations to no better than 9 arcsec.

Bouvard's rejection of the older observations was not acceptable to all astronomers at that time. And alas! By 1825 the discrepancy between Bouvard's predicted positions of Uranus and the observed positions began to grow. By 1832 Uranus was nearly 30 arcsec from its predicted position, and even though this is only about a sixtieth of the Moon's angular diameter in our skies, it was *far* too large to be attributed to the observational uncertainties, which were only a few arcsec.

Various solutions were suggested for the problem of Uranus. Among these was that Newton's Law of Gravity did not apply beyond Saturn. Another suggestion was that beyond Uranus there were one or more unknown bodies, massive enough to be planets.

A planet beyond Uranus?

A planet beyond Uranus could, in principle, solve the problem of Uranus's orbit through its gravitational attraction on Uranus. A fair assumption was that it orbited the Sun in the same direction as all the other planets, the prograde direction. Suppose that at a certain time it orbited ahead of Uranus. In this case it would pull Uranus forward. Later, in accord with Kepler's third law, Uranus, being in the smaller orbit, will have moved ahead of the other planet, which would then pull Uranus back. The details are rather complicated – all I want you to realise is that a trans-Uranian planet will influence the orbit of Uranus. The devil is in the detail, the actual effect depending on the mass and orbit of the unknown planet. (These details are beyond the scope of this book.)

One of the first documented suggestions that there was a trans-Uranian planet came from the English clergyman and amateur astronomer, the Reverend Doctor Thomas John Hussey (born 1797), in a letter he sent in 1834 to the then Plumian Professor of Astronomy at Cambridge University, George Biddell Airy (1801–1892). Hussey also suggested that he, Hussey, should undertake a search, but Airy discouraged him, perhaps because of Airy's belief that the solution lay in modifying Newton's Law of Gravity. In 1842 the German astronomer Friedrich Wilhelm Bessel (1784–1846) gave to a research student the task of predicting the mass and orbit of a trans-Uranian planet, but the student died, and Bessel became too ill to take up the challenge. Then he too died.

FIGURE 2.3 John Couch Adams in an engraving published in 1851. He made several predictions of the mass and orbit of a trans-Uranian planet. (Institute of Astronomy, Cambridge, UK, by permission)

Adams and Le Verrier

One of the central figures in the discovery of Neptune is John Couch Adams (1819–1892), the son of a Cornish farmer (Figure 2.3). His talent for mathematics became apparent at an early age, and in 1839 he gained admission to Cambridge University to study mathematics. He graduated brilliantly in 1843 by which time he had become interested in the problem of Uranus and was convinced that the solution lay in the existence of a trans-Uranian planet. James Challis (1803–1882), Plumian Professor of Astronomy after Airy became Astronomer Royal in 1835, gave Adams some encouragement and by October 1843 Adams had shown that it was *possible* that a trans-Uranian planet could account for the observed motion of Uranus. By 1845 Adams had calculated a mass and an orbit for such a planet, and showed that it could account for all of the known observations of Uranus including the pre-discovery observations. The main assumptions that Adams had made were that the semimajor axis of the planet was 38.4 AU in accord with a version of the Titius-Bode rule and that the orbital inclination was negligibly small. He deduced an orbital eccentricity of 0.16 and a planetary mass of 49.9 M_E.

Adams showed his calculations to Challis and then left Cambridge to visit his family in Cornwall, leaving Challis with a position for the new planet for 30 September 1845. Adams had also pointed out that the planet should have a disc large enough to be seen with the biggest telescope at Cambridge, the Northumberland refracting telescope, which had a main lens 11.6 inches (295 mm) in diameter. Challis made no search.

On his way to Cornwall Adams visited Greenwich to present Airy with a summary of his calculations, but Airy was in Paris. On his return from Cornwall Adams again called at Greenwich, but Airy was out and so Adams left him a summary of his results. Adams returned to Airy's house later in the day but was turned away by the butler. It seems that Adams felt slighted.

Adams later heard from Airy that he was not enthused by the summary that Adams had left, and this probably contributed to Adams's reduced interest in a trans-Uranian planet. There are at least three possible reasons for Airy's cool response: he was probably still convinced that a modification of Newton's Law of Gravity was the correct explanation of Uranus's errant behaviour; he was not convinced that a calculation along Adams's lines could be performed with sufficient accuracy, indeed he was more inclined towards practical work; and he distrusted youth.

In December 1845 a research paper reached Adams that had been presented to the French Academy of Sciences a few weeks earlier, on 10 November. It contained a very thorough analysis of the motion of Uranus based on the observations made between 1781 and 1845. It, incidentally, discredited the work of Bouvard, showing it to contain many errors. But the paper did not reconcile the motion of Uranus with its calculated orbit. For example, the orbit based on the observations made between 1790 and 1820 placed Uranus over 40 arcsec from its actual position at the 1845 opposition (at opposition a body, in this case Uranus, lies in the opposite direction to the Sun as viewed from the Earth). The paper concluded that the discrepancy

> 'can be attributed to outside causes whose effect I will evaluate in a second Memoir'.

Airy regarded this paper as

> 'a new and most important investigation'

and that

> 'the theory of Uranus was now, for the first time, placed on a satisfactory foundation'.

The author of the paper was Urbain Jean Joseph Le Verrier (1811–1877).

Le Verrier, son of a civil servant, was born in Normandy (Figure 2.4). He showed some early talent for science and gained admission to the École Polytechnique in Paris in 1831. He seemed set for a career in chemistry when a good position in astronomy became available to him at the École, which he took up; in those days there were few barriers between the different scientific disciplines.

At first Le Verrier spent a good deal of time on studies of the orbits of comets, a subject which had also occupied Adams. But in the summer of 1845, following on from a small amount of earlier work, he was encouraged to concentrate on Uranus. His paper of 10 November 1845 was the first fruit.

On 1 June 1846 Le Verrier presented his second paper on Uranus to the French Academy of Sciences. In it he presented calculations of the mass and orbit of a trans-Uranian planet, and he gave its position in the sky for 1 January 1847, with an estimated accuracy of about 10° for its direction in the sky, as seen from the Earth. The main

FIGURE 2.4 Urbain Jean Joseph Le Verrier, who also made several predictions of the mass and orbit of a trans-Uranian planet. (Bibliothèque de l'Observatoire de Paris, by permission)

assumptions, like those of Adams, were that the orbit had a low inclination and that the semimajor axis was in the range 35–38 AU (again in accord with a version of the Titius-Bode rule).

When Airy received this second paper in late June he could not

> 'sufficiently express the feeling of delight and satisfaction which I received from it'.

He probably noticed that Adams's and Le Verrier's position for the new planet agreed to within 1°. Nevertheless, when he wrote to Le Verrier on 26 June he did not mention Adams.

I find it difficult to understand Airy's behaviour. It could be that he now realised that with two solutions in such agreement there might be a trans-Uranian planet after all, and that if Le Verrier was kept ignorant of Adams's work then perhaps Le Verrier would not press French astronomers to search for the planet; the discovery might then fall to Britain. However, this does not explain why Airy was so *encouraging* to Le Verrier. Was he a sycophant?

The search is on (and off)

It was certainly not long before Airy pressed Challis and others to set up a systematic search for a trans-Uranian planet, using the Northumberland telescope at Cambridge, with a main lens 11.6 inches (295 mm) in diameter. Airy did not want this to be done at Greenwich, perhaps because the largest telescope had a main lens only 6.7 inches (170 mm) in diameter, and perhaps because he did not want to disturb the large number of routine observations being made there. Moreover, Airy had been responsible for setting up the Northumberland telescope.

Challis does not seem to have been particularly keen to embark on the search. This might have been because of personal and professional conflicts with Airy, coupled with Challis's scepticism about the existence of the planet in spite of his erstwhile encouragement of Adams.

Airy became concerned at the lack of action, and pressed Challis to start. On 29 July 1846 Challis began his search. He started in a region of the sky where Adams had recently told him the planet should be found. The method Challis adopted was to record the position of every star and check each against star charts; a 'star' where none existed on the chart might be the planet. At that time this was an obvious strategy because most observational work was on stars and asteroids, objects which appear as tiny discs of light corresponding to the optical limitations of the telescope. However, Adams had urged Challis to use a high magnification and search for a larger disc. High magnification carries the penalty of a smaller field of view, but has the huge advantage of being at once able to distinguish a planetary disc from the many stars in the field. Challis ignored Adams's advice.

By the end of August 1846 Adams had further refined his calculations and had revised the mass and orbit of the planet. By then Challis had observed the area of sky in which Adams had previously predicted that the planet would lie. Challis had not checked all of his observations, but none of those he had checked had revealed a 'star' unrecorded on the star charts. It seems that Adams's new positions, which did not differ greatly from the earlier ones, failed to inspire Challis to complete the checks.

On 31 August 1846 Le Verrier published his third paper on a trans-Uranian planet, in which he presented a mass and an orbit with even greater accuracy than before. He had also recalculated the position of the planet for 1 January 1847. The new mass was 35.8 M_E, and the new orbit, shown in Figure 2.5, had a semimajor axis of 36.154 AU and an eccentricity of 0.17. This figure also shows the August 1846 orbit of Adams, for a planet with a mass 49.9 M_E. The two 1846 predicted positions in Figure 2.5 were not far apart on the sky as seen from the Earth. Le Verrier had also predicted that the planet should have a disc large enough to be seen as such in Earth-based telescopes. In particular, he had estimated that at the predicted opposition of the planet on 19 August 1846 its angular diameter as seen from the Earth would be 3.3 arcsec. This size estimate, and a similar one by Adams, was obtained from the predicted masses of the planet by making reasonable assumptions about its density. The angular diameter would not change measurably for several months around opposition.

Le Verrier had even less success persuading French astronomers to search for the planet than Adams had received in England, where, apart from Challis, there were certainly no more than one or two others making a serious search, and perhaps no others at all.

Le Verrier turned to Germany. On 23 September 1846 a letter from him reached Johann Gottfried Galle ('gal-ay', 1812–1910) who was an assistant at the Berlin Observatory. Galle quickly persuaded the Observatory director, Johann Franz Encke (1791–1865), to allow

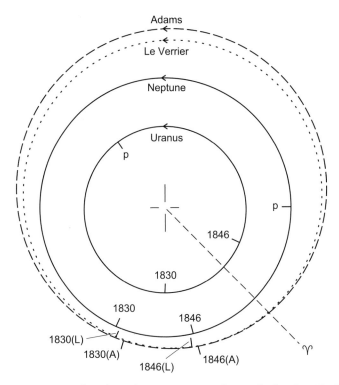

FIGURE 2.5 The orbits of a trans-Uranian planet calculated in 1846 by Le Verrier and Adams. Its predicted positions in August 1846 are labelled L for Le Verrier and A for Adams. The orbits and corresponding positions of Uranus and Neptune are also shown for the Neptune discovery date, 23 September 1846, though Neptune only moves around its orbit at 0.18° per month. The 1830 positions are also shown. Perihelia are marked by 'p'. The viewpoint is perpendicular to the orbit of Neptune, though the orbital inclinations are all so small that the shapes of the other orbits are unaffected on the scale of this figure. The dashed line marked γ is a reference direction in the Solar System that points to a particular location among the distant stars – the first point of Aries.

him to search for the planet, and that same night he began, assisted by a young student astronomer Heinrich Louis d'Arrest (1822–1875). Using the Fraunhofer Refractor, with a main lens diameter of 9.0 inches (229 mm), Galle searched the sky in a region where, using Le Verrier's orbit, he had calculated the planet should lie that month. At Le Verrier's suggestion he searched for a disc, but found none. But they then checked the stars they could see against a star chart, Galle at the telescope, d'Arrest with a new (1845) chart of that area of sky. Only a few had been checked when d'Arrest exclaimed that Galle had called out a star that was *not* on the chart.

They watched the object until it set at 02:30. Was it the planet? They picked it up the following night; it had moved at roughly the right rate against the stellar background, 3 arcsec per hour. Careful scrutiny at an angular magnification of 320 showed that it had a disc only a little less than the 3.3 arcsec predicted by Le Verrier. The trans-Uranian planet had been found, slightly less than a degree from Le Verrier's predicted position, and a little over two degrees from Adams's position.

The aftermath

It was some months before the planet's name was decided: Neptune, in Roman mythology the son of Saturn and the ruler of the ocean deeps.

By an ironic coincidence Adams was in Germany at the time of the discovery, and therefore heard of it several days before the news reached Britain. It was announced in The Times of London on 1 October 1846, in a letter to the editor from John Russell Hind (1823–1895), director of the Bishop's Observatory in Regent's Park, London.

I wonder what Challis felt like on that October day. Just two days earlier, on 29 September, he had seen Le Verrier's 31 August paper and decided at once to search for a disc. That night he saw an object that seemed to have a disc but decided to check it another day. On 1 October, after he had read of the discovery, he did the check, and it *was* Neptune. But it was then too late to claim an independent discovery. More agonising yet, Challis found that he had recorded Neptune on 30 July, but in checking that part of the sky against a star chart he had stopped just ten stars short of the planet. He had recorded the planet again on 4 August, but again that area of sky had not been compared against a star chart.

In early October Challis made public his months of searching for a trans-Uranian planet, but curiously did not mention Adams. It was John Herschel (son of William), one of the few other people who knew of Adams's work, who publicised it on 3 October. This at once raised French suspicion that British astronomers were trying to muscle in on a French discovery. Why, it was asked, had neither Challis, nor Airy, when he had written to Le Verrier in June, mentioned Adams? The French press became fairly hysterical, and the contemporary political tensions that existed between Britain and France were drawn into the issue.

Gradually, the storm subsided. Adams and Le Verrier met, became friends, and remained so until Le Verrier's death in 1877. Within a few years of 1846 it became widely acknowledged that Adams and Le Verrier were *co-predictors* of Neptune, though Galle and d'Arrest are regarded as *sole discoverers*.

Figure 2.5 also shows the *actual* orbit of Neptune. You can see that, away from the discovery region, it differs considerably from the orbits calculated by Adams and Le Verrier, mainly because the actual orbit has a lower eccentricity and is smaller than the two predicted orbits. Moreover, we now know that Neptune's actual mass is only 17.15 M_E, about half of Le Verrier's prediction, and about a third that of Adams's. This has led one or two astronomers to suggest that the discovery of Neptune was a happy accident.

This is *not* so. The orbits of Adams and Le Verrier place their predicted planets fairly close to Neptune throughout the first few decades of the nineteenth century, as Figure 2.5 shows. You can also see from Figure 2.5 that Uranus was fairly close to Neptune during this period, and it was from Neptune's effect on Uranus that the orbit of Neptune was deduced. It is therefore no accident that Adams and Le Verrier place their planets in roughly the right *direction* during this period. The larger orbits of their planets than that of Neptune, around 38 AU versus 30.20 AU, are because of their use of versions of the Titius-Bode rule. The larger orbits also led to their predicted masses for Neptune being too large.

From the actual orbit of Neptune, and the discrepancies in the orbit of Uranus, the first estimate of the true mass of Neptune was made by others, 22.2 M_E in February 1847, soon revised to 16.8 M_E, which is not far from the modern value. That a single mass and a particular orbit could account for the discrepancies in Uranus's position (to within the then observational uncertainties of a few arcsec) showed that the source of these discrepancies had been found.

Many pre-discovery observations of Neptune are now known. The earliest is by Galileo Galilei (1564–1642), in Italy. In December 1612 and January 1613, whilst he was observing Jupiter and its newly discovered satellites, he recorded a 'star' that moved with respect to the other stars. I wonder what the effect would have been on the history of astronomy had he correctly identified the moving star as a planet?

Some consolation for British astronomy came on 10 October 1846 when William Lassell (1799–1880) suspected that he had discovered a satellite of Neptune using a recently completed reflecting telescope at Liverpool, with a main mirror diameter of 24 inches (610 mm), the biggest telescope in Britain at that time. By July 1847 he had confirmed its existence. For a long time it was known as Neptune's satellite, but after the discovery of a second satellite (Nereid) in 1949, a name that had been first suggested in 1880 was universally adopted. The name is Triton, the Greek god of the sea, and son of the Greek god Poseidon, in Roman mythology, Neptune. Its orbit enabled the mass of Neptune to be calculated far more accurately than from Neptune's effect on Uranus. Over the next few years various values were calculated, homing in on values like 17.73 M_E and 17.16 M_E, the latter being extremely close to the modern value of 17.15 M_E. You will see in Section 4.4 that there are reasons to believe that Triton resembles Pluto.

2.3 THE DISCOVERY OF PLUTO

Following the discovery of Neptune, observations of Uranus and Neptune accumulated, including more pre-discovery observations of both planets. These observations enabled their orbits to be calculated more precisely. Alas! Within a few decades it became clear that their motions could no longer be deduced from the gravitational pulls of the other known bodies in the Solar System. A trans-Neptunian planet was suspected. After some speculations and false starts by others, the first serious search for such a planet was made in 1877 by the American astronomer David Peck Todd (1855–1939), at the US Naval Observatory. He based his search on the discrepant motion of Uranus, the orbit of which, at that time, was known more accurately than that of Neptune, because it had been observed for longer than Neptune, indeed for longer than its orbital period of 84 years. The gravitational effect of Neptune was, however, included in predicting Uranus's motion. He deduced that a trans-Neptunian planet in a 52 AU orbit existed, and predicted its position in the sky. But his six month search failed to reveal such a planet.

In 1878 the French astronomer, and one-time assistant of Le Verrier, Camille Flammarion (1842–1925) predicted the existence of two trans-Neptunian planets, one at 100 AU from the Sun, the other at 300 AU. His prediction arose from his observation that several comets seemed to have aphelia at these sorts of distances. These planets were not found. Nor were those of the Danish astronomer Hans Emile Lau (1879–1918) who, like Todd, used the discrepant motion of Uranus, but predicted the existence of two planets, one at 46.6 AU, the other at 70.7 AU. He assigned masses, 9 M_E and 47.2 M_E respectively. Others entered the fray, equally unsuccessfully.

And then the whole issue was seemingly put to rest by the French astronomer Jean Baptiste Aimable Gaillot (1834–1921). From the observations of Uranus and Neptune he calculated revised orbits, and concluded that there were *no* discrepancies – his orbits fitted the observations! Most astronomers accepted this conclusion, but among the few who did not were the American astronomers William Henry Pickering (1858–1938) of Harvard, and Percival Lowell (1855–1916), a wealthy Bostonian (Figure 2.6).

FIGURE 2.6 Left: William Henry Pickering in 1909. Right: Percival Lowell. (Left, US Library of Congress, Right, Lowell Observatory, by permission)

William Henry Pickering

Pickering called his trans-Neptunian planet, Planet O. He calculated its mass and orbit from the discrepancies in the well-known orbit of Uranus, and in 1908 came up with a mass of 2 M_E and an orbit with a semimajor axis of 51.9 AU. In 1911 Pickering came up with three more trans-Neptunian planets additional to Planet O. These he named Planets P, Q and R, and announced that Planet P probably existed. He subsequently modified the masses and orbits of Planets O and P, as shown in Table 2.1. The 1928 orbit of Planet O, the final one, had an eccentricity such that it crossed the orbit of Neptune. For the final

FIGURE 2.6 (*cont.*)

orbit of Planet P, after the discovery of Pluto in 1930, he calculated an orbital eccentricity of 0.265, and an inclination of 37°, not very different from his 1911 orbit.

In 1928 Pickering had predicted the existence of three more trans-Neptunian planets, which he called S, T and U. Their masses and orbits were also revised after the discovery of Pluto.

Pickering's planets comprise quite a menagerie, but it seems to have been Planet O that he encouraged others to search for, and indeed he searched for it himself. But the searches were, on the whole, rather perfunctory. Planet O was never found, nor were Planets P, Q, R, S, T and U. His part in the discovery of Pluto is considerably smaller than that played by Percival Lowell.

Table 2.1 *Masses and orbits of Pickering's Planets O and P.*

Year	Planet O mass (M_E)	Planet O a (AU)	Planet P mass (M_E)	Planet P a (AU)
1908	2.0	51.9	–	–
1911	–	–	???	123
1919	2.0	55.1	–	–
1928	0.75	???	20	67.7
1931	–	–	50	75.5

Percival Lowell

Percival Lowell was born into a wealthy New England family. His education included undergraduate years at Harvard University, where he studied mathematical astronomy. Afterwards he made a career in business and in the diplomatic service. In 1893 he returned to astronomy, and began to build an observatory on Mars Hill in Flagstaff, Arizona, primarily for the study of Mars. Lowell was one of the minority of astronomers who, at that time, thought that Mars bore canals built by a technologically accomplished civilisation, but that's another story! The observatory was also used, from 1905 onwards, to search for Planet X, Lowell's name for his trans-Neptunian planet. His (photographic) search lasted from 1905 until his death in 1916.

Lowell followed the method used by Adams and Le Verrier i.e. the discrepant motion of a known planet, and again Uranus was chosen because of its better known orbit than that of Neptune. But whereas in the years before Neptune's discovery the position of Uranus was up to 133 arcsec from its predicted position, with the gravitational influence of Neptune now included, the discrepancy early in the twentieth century never exceeded 4.5 arcsec. This smaller discrepancy reduced the accuracy with which the orbit and mass of Planet X could be calculated.

Lowell asserted that the semimajor axis, a, of Planet X's orbit would be close to 47.5 AU, in accord with the Titius-Bode rule, which

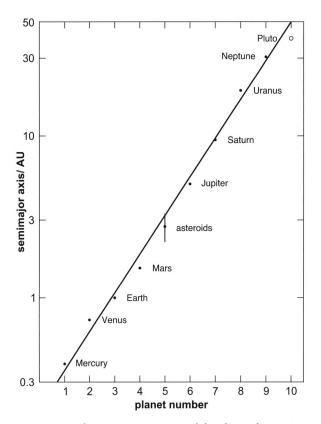

FIGURE 2.7 The semimajor axes, a, of the planets known up to 1930, versus their order n from the Sun: $n = 1$ Mercury, 2 Venus, 3 Earth, 4 Mars, 5 the asteroids, 6 Jupiter, 7 Saturn, 8 Uranus, 9 Neptune. Pluto has been added, though it was not discovered until 1930.

Figure 2.7 shows with Neptune included. You can see that Neptune fits the rule rather well, which justified its use to predict the value of a for Planet X.

In 1913, with $a = 47.5$ AU, Lowell predicted that Planet X would have a mass of 6.6 M_E. Subsequently, in 1914, he reduced a to either 43.0 AU or 44.7 AU. But in spite of these and subsequent refinements to the orbit of Planet X, and searches spanning 10 years, he failed to discover the planet. This failure was the biggest disappointment of his life, though sadly he had recorded images of Pluto, as a faint speck of

FIGURE 2.8 The Pluto discovery telescope, a 13-inch (330-mm)
diameter refractor, specially designed for astrophotography – an
astrograph. (Lowell Observatory, by permission) (See plate section for
colour version.)

light, on two photographic plates obtained at the Lowell Observatory
in 1915; Lowell was looking for a much brighter object.

Clyde Tombaugh

Percival Lowell's death in 1916 put a stop to searches for Planet X at
the Lowell Observatory. His will ensured the survival of the observatory and requested that the search for Planet X continue. Unfortunately, there was a dispute over the will involving Lowell's widow that
prevented the Observatory operating. Fortunately, before the matter
was resolved, Lowell's brother Abbott (1856–1943) donated money in
1925 to purchase a telescope to restart the search for Planet X. This
was to be a refractor with a main lens 13 inches (330 mm) diameter,
its large size making it good for detecting faint objects. It was to be
equipped with a camera that, with the telescope, would have a large
field of view, about 800 times the area of the full Moon – good for
planet searching (Figure 2.8).

In 1928, with the imminent arrival of the telescope, the director of the Lowell Observatory since 1916, Vesto Melvin Slipher

(1875–1969), decided that it was time to acquire an assistant to carry out the search for Planet X. He was in correspondence with a farm boy from Kansas, self-taught in astronomy, about some drawings of Mars and Jupiter that the young man had sent him, drawn using his own home made telescope. Slipher had been impressed, and within a few months invited the young man to join the staff at Lowell Observatory. His name was Clyde William Tombaugh (1906–1997).

Tombaugh arrived at Flagstaff Observatory in February 1929, not knowing that his job would be to use the new telescope to search for Planet X. He soon found out, and on 6 April 1929 his search began. In the intervening years since Lowell's death, William Pickering and his assistant Milton Lasell Humason (1891–1972) had carried out a search for a trans-Neptunian planet using photographs obtained at the Mount Wilson Observatory, but without success. I wonder if this was a challenge to Tombaugh, or a cause of dismay? I suspect the former.

Slipher had told Tombaugh that Planet X would be at least 15 times fainter than Neptune, and that long exposures, of an hour or more, were necessary to detect such a faint object. Tombaugh was also told that each area of sky was to be photographed at least twice, with several nights separating each exposure. This was because planets move slowly against the fixed pattern of the much more distant stars. The sky also needed to be dark, requiring very clear skies and no Moon.

The telescope, like the great majority of telescopes in those days, was on what is called an equatorial mount. The essential feature is that one axis of rotation, called the polar axis, is set up parallel to the Earth's rotation axis. This means that once the telescope is pointed at the desired area of the sky, the Earth's rotation, which causes celestial objects to rise in the East and set in the West, can be cancelled out by rotating the telescope around the polar axis at the same rate at which the sky appears to rotate, but in the opposite direction. In this way a celestial object is held in the field of view in a fixed position. However, as I well know from my own experience, equatorial mounts are not perfect, requiring small corrections every now and then, to keep a

FIGURE 2.9 Clyde Tombaugh, at Lowell Observatory in about 1931, carrying a photographic plate. (Lowell Observatory, by permission)

guide star fixed in the field of view (either of the main telescope or of a smaller 'finder' attached to it). Therefore, Tombaugh could not slip off for a hot drink during the hour or so exposures; constant vigilance was essential.

After a few weeks of searching, Slipher added to Tombaugh's duties by getting him to inspect the 356 × 432 mm photographic plates for evidence of Planet X. This was done using a device called a blink comparator, in which two plates obtained several nights apart were viewed alternately by manual switching to see if any 'star' had moved with respect to the other stars. The moving object would certainly not be a star. It might be an asteroid, or a distant comet before it developed its tail, or Planet X. Each plate had images of up to 900 000 stars, and it required a whole day to examine just a few square centimetres of each pair. This was a very tedious task. Moreover, attention could not lapse for a moment, lest Planet X be missed! And all this after a night obtaining the exposures.

Tombaugh did not concentrate on the position of Planet X predicted by Lowell. Instead, after a few months, he realised that the best region of sky to search was around the opposition point, i.e. the direction opposite from the Sun as seen from the Earth. There are three reasons.

- Planet X would be moving at its greatest rate with respect to the distant background stars, because the Earth is then moving at right angles to the line of sight to Pluto (Figure 2.10).
- Planet X is closer to the Earth than at any other time until the next opposition (which occurs a little *over* a year later, because Planet X would have moved slightly around its orbit).
- Planet X is highest in the sky at midnight, and so it can be observed in a dark sky for the longest time.

Tombaugh therefore concentrated on the region around opposition. As the Earth moves around its orbit the opposition region sweeps out a band across the sky. The opposition point for a planet in an orbit with zero inclination is, by definition, in the ecliptic plane (Section 1.2), and as the inclinations of the then known planetary orbits beyond the Earth are only a few degrees, Tombaugh restricted his search to near this plane.

Success!

It was on 18 February 1930 that Tombaugh examined the plates he had taken in the region of the star Delta Geminorum between 21 and 29 January. He noticed a faint dot of light jumping a little when using

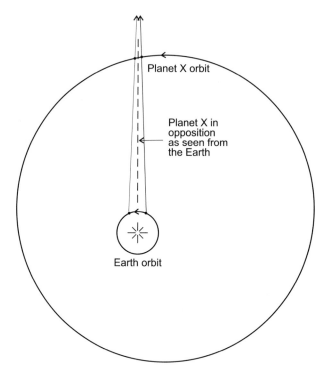

FIGURE 2.10 Planet X near the opposition point as seen from Earth (not to scale).

the blink comparator to switch between the plates exposed on 23 and 29 January (Figure 2.11). It only jumped a few millimetres across the plate, but this was good, because an asteroid, being much closer to the Sun and therefore moving more quickly around its orbit, would have moved much more than a few millimetres in the six days that separated the exposure of the plates.

Though Tombaugh thought he had probably discovered a planet he was a careful astronomer. He therefore examined plates he had taken on the same nights with a 5-inch (127-mm) refractor mounted on the 13-inch refractor. If, in both telescopes, the planet was in the same position with respect to the other stars, some flaw in the larger telescope and/or its camera could be ruled out. The planet *was* in the

DISCOVERY OF THE PLANET PLUTO

January 23, 1930

January 29, 1930

FIGURE 2.11 Portions of the two discovery plates of Pluto: Left 23 January 1930, Right 29 January. (Lowell Observatory, by permission)

same position in the plates from each telescope. He then checked the other plates he had taken of the Delta Geminorum region. The planet was there in all of them, it was real! He at once told Slipher. It was still 18 February.

It was not until 13 March that the discovery was made public, a date selected because it was the 75th anniversary of Lowell's birth and the 149th anniversary of the discovery of Uranus. It was also the case that by then further observations had put the existence of the planet beyond doubt.

The new planet was not named until May 1930. Many names were suggested, but the staff of Lowell Observatory favoured Pluto, which had been suggested by an eleven year old girl from Oxford, UK, Venetia Burney, later Mrs Venetia Phair (1918–2009). Pluto is the Greek god of the underworld, very appropriate for a planet so far from the Sun. Also, PL was like Percival Lowell's monogram, P_L. By the end of May 1830 the name was accepted by the American Astronomical Society and the UK's Royal Astronomical Society.

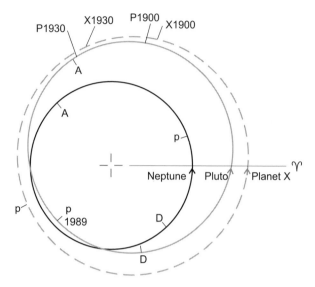

FIGURE 2.12 Orbit 1914B of Percival Lowell's Planet X, along with the orbits of Neptune and Pluto. The viewpoint is perpendicular to the ecliptic plane (though Neptune's orbit only has the low inclination of 1.77°). Pluto's orbit has an inclination of 17.1°, and would appear slightly more elliptical if viewed face-on. It dips below the ecliptic plane between D and A. The inclination of Orbit 1914B is 10°, but the crossing points of the ecliptic plane are unspecified. Perihelia are denoted by p.

2.4 PLUTO SURPRISES

Figure 2.12 shows the orbit of Pluto (and Neptune), and one of two of Lowell's orbits for Planet X from 1914, called 1914B. The associated mass of Planet X is 6.6 M_E, the semimajor axis, $a = 43.0$ AU, the orbital inclination, $i = 10°$ and the eccentricity, $e = 0.202$. Note that his prediction of an orbit does not give a unique set of orbital elements and mass, but two sets, in this case the main differences being that Planet X could be in either of two positions in its orbit, separated by about 180°, and that the masses were 6.6 M_E in the orbit shown in Figure 2.12, and 7.6 M_E in the other orbit. (See Box 2.2.)

By mid-1930 further observations of Pluto, including pre-discovery photographs, resulted in Pluto's orbit being known rather well – in the 1930s a semimajor axis of 39.5 AU, an eccentricity of

0.248, and an inclination of 17.1° (in Table 1.1 the values are for 2009, and are slightly different in the cases of *a* and *e*). There were two surprises here. First, the eccentricity is larger than that of any other planet (Table 1.1), so much so that around perihelion it is closer to the Sun than Neptune, as you can see in Figure 2.12. You might think that this means the two orbits intersect. This is not so; I'll return to this point in Section 2.5. Second, the inclination is higher than for any other planet (Table 1.1). Thus, within a few months of its discovery it was realised that the orbit of Pluto is strikingly different from the other planets' orbits.

Furthermore, the semimajor axis of the 1914B orbit of Planet X, 43.0 AU is significantly larger than that of Pluto, a difference also clear in Figure 2.12. Perhaps this is due, at least in part, to Lowell's tendency to avoid having his value for Planet X much smaller than that predicted by the version of the Titius-Bode rule that he was using. In Figure 2.7, you can see that Pluto plots well below the line that the other planets are close to. Let's set the Titius-Bode rule aside.

Pluto is nearly ten times fainter than Lowell had predicted for Planet X, a prediction based on the mass needed to perturb Uranus plus an assumed surface reflectance. This could indicate that Pluto is smaller than expected, and indeed this was found to be the case; even the biggest telescopes in the 1930s could not image Pluto as a disc, but only as a tiny diffraction limited blurred disc. The mass of Pluto could not be anywhere near as large as the 6.6 M_E required by Lowell to explain the discrepant motions of Uranus. Therefore, even though Pluto was discovered not far from Lowell's predicted position (Figure 2.12), it seems that this is a coincidence – Pluto is *not* Planet X.

This conclusion led to renewed searches for a trans-Neptunian planet over the following several decades, during which time Pluto's calculated mass became even smaller (Chapter 3). No planet was found.

So, what caused the discrepancies in Uranus's orbit? The answer to this question came in 1993, when the American astronomer Erland Myles Standish announced that, in fact, there *were* no discrepancies!

Table 2.2 *Pickering's Planet O (1919), Lowell's Planet X (1914B) and Pluto.*

	Mass (M_E)	a (AU)	e	i (°)
Planet O (1919)	2.0	55.1	0.31	15
Planet X (1914B)	6.6	43.0	0.202	10
Pluto (1930)	$< \sim 1$	39.5	0.248	17.1
Pluto (mid 2009)	0.00218	39.64	0.251	17.14

BOX 2.2 SOLUTION AMBIGUITY, A SIMPLE EXAMPLE (OPTIONAL)

There are many instances, as well as in orbital determination, where more than one solution is possible. A simple example is in the determination of the square root of a number, i.e. what number multiplied by itself gives the number in question. If I were to ask you what is the square root of 4, you would very probably answer 2. This is correct, but it's not the complete answer. The alternative answer is −2. When −2 is multiplied by itself the two minuses become a plus, so the answer is again 4. Thus the complete answer to the question 'what is the square root of 4?' is '+2 or −2'.

He used data from the flybys of the giant planets by the NASA spacecraft Voyagers 1 and 2. The gravitational effect of Neptune on Voyager 2 led to a decrease of 0.5% in Neptune's mass, which enabled Standish to come to his astonishing conclusion. As Shakespeare might have said, 'much ado about nothing'.

Pickering's Planet O

I cannot leave the pre-discovery predictions of Pluto's mass and orbit without a brief mention of Pickering's predictions. Table 2.2 shows his 1919 prediction for Planet O along with that of Lowell's 1914B orbit for Planet X, and the 1930 and present values for Pluto.

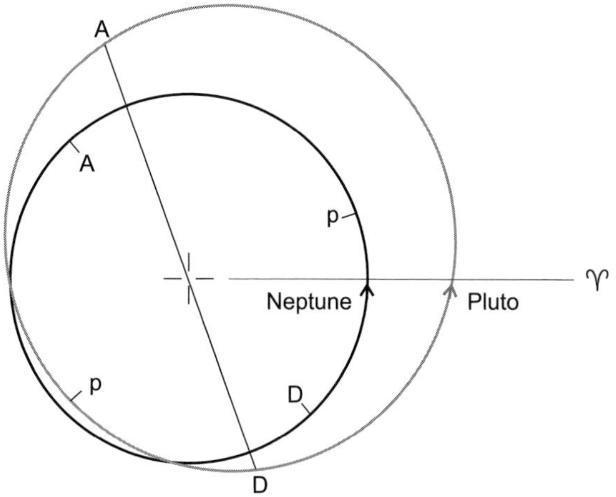

FIGURE 2.13 The orbits of Pluto and Neptune. The points marked A are the ascending nodes, the points where the orbits cross from below the ecliptic plane to above it, and D marks the descending nodes.

You can see that Pickering's 1919 prediction of the mass of his Planet O is closer to the actual value for Pluto than Lowell's value. Moreover, this is also the case for his prediction for i. In 1928 he reduced the mass of Planet O to just 0.75 M_E, far closer to the actual value than Lowell's value.

The trouble with Pickering's work is that he made many quite different predictions, with various numbers of planets. Even after Pluto's discovery, Pickering preferred three planets rather than a single planet, Planet O. Planet O is no more Pluto than Planet X is, indeed, rather less so.

2.5 WHY NO COLLISIONS WITH NEPTUNE?

I have stated that, in spite of what a cursory glance at Figure 2.12 might suggest, the orbits of Pluto and Neptune do not intersect. Figure 2.13 shows just these two orbits, each of which lies below the ecliptic plane between D and A. At D the orbits of Pluto and Neptune cross the ecliptic plane from above the plane to below it, and at A they cross from below to above; these are called the descending and ascending

FIGURE 2.14 The side view of the orbits of Pluto and Neptune with respect to the ecliptic plane. You might think that the inclination of Pluto's orbit with respect to the plane of Neptune's orbit is $(17.1° − 1.77°) = 15.3°$. However, this would only be the case if the two ascending nodes, A, in Figure 2.13 lay on the same straight line to the Sun. That this is not quite the case means that the actual value of the difference in orbital inclinations is slightly different.

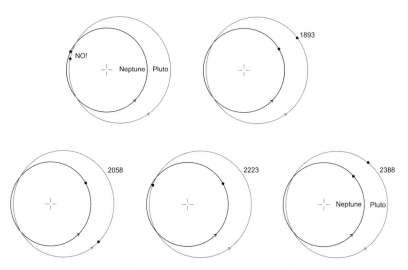

FIGURE 2.15 The 3:2 mean motion resonance of Pluto and Neptune prevent them from getting close to each other. The top left panel shows a configuration that can never occur. The remaining panels show the configurations in each of the four years shown, in none of which are Pluto and Neptune close to each other. Each panel is one orbit of Neptune later than the one before. The large orbital inclination of Pluto leads to even greater separations.

nodes respectively. Figure 2.14 shows the side view of each orbit with respect to the ecliptic plane (at a smaller scale than Figure 2.13). From Figures 2.13 and 2.14 you can see that the two orbits are always well separated.

But this is only part of the story. Pluto and Neptune cannot collide, but they could, in principle, get close enough in the arc of Pluto's orbit around its perihelion for the massive planet Neptune to destabilize the orbit of Pluto. Why has this not happened? The answer is that for every three orbits of Neptune, Pluto orbits the Sun twice. In other words the ratio of the period of Pluto's orbit to that of Neptune is 3:2 (in the second half of 2009, 249.44 years/165.84 years, which is 1.504, a ratio of 3.008:2). This is an example of a mean motion resonance, in this case the 3:2 mean motion resonance. This resonance is stable, in that the orbital elements of Pluto are confined to small oscillations, and that there are only slight departures from 3:2, which is the average over the age of the Solar System.

Figure 2.15 shows how this works. You can see that Pluto and Neptune never get close around Pluto's perihelion arc. They could get close if, for example, Pluto and Neptune could be at the positions in the top left panel. But this never happens. There are many other cases of mean motion resonances in the Solar System, not always 3:2, but, for example 2:1, where for every two orbits of a body, another body completes just one orbit.

In the next chapter the focus is on Pluto as a world, rather than on its orbit.

3 Pluto: a diminishing world

As soon as Pluto was discovered, astronomers were eager to learn as much as possible about this remote world. What type of body was it that lurked at the outer edge of the Solar System? The most fundamental properties are size and mass. These give the mean density by dividing the mass by the volume; the mean density in turn constrains Pluto's composition.

3.1 PLUTO'S SIZE

If Pluto could be seen as a disc then, with its distance known, its size could be estimated from its measured angular diameter, as described in Section 1.6. You might think that with a sufficiently large telescope a disc would have been seen. For telescopes at the Earth's surface this is not the case. There are two reasons for this, given in Box 2.1, reasons that I slightly enlarge upon here.

First, there is the intrinsic limit of the optics (the diffraction limit), the larger the main lens or mirror, and/or the shorter the wavelengths detected, the smaller the fuzzy disc image of a point of light and the better the telescope's resolution. Visible light covers the wavelength range of about 0.38 millionths of a metre (a micrometre), to about 0.75 micrometres. The human eye is most sensitive at about 0.55 micrometres, which we see as green. At this wavelength, a telescope with a perfect main lens or mirror a metre in diameter produces a fuzzy disc such that two points of light separated by about 0.14 arcsec (each imaged as a fuzzy disc), could just be distinguished. The angular resolution is then about 0.14 arcsec.

Second, turbulence in the Earth's atmosphere blurs the image further. This blurring limits the angular resolution at visible wavelengths to about 0.4 arcsec under very steady skies, which corresponds to a perfect lens/mirror only about 300 mm in diameter. Sometimes the 'seeing', as it's called, is better than this, but usually it's worse. The big advantage in going to large sizes is that more light is collected and therefore fainter objects can be seen with the eye, and shorter exposure times can be used to record them, photographically or electronically.

In 1930 there were plenty of telescopes larger than 300 mm; the largest of all was the 100-inch (2.54-metre) reflecting telescope on Mount Wilson, near Los Angeles. But, as you have learned, beyond about 300 mm diameter size doesn't matter when it comes to detecting discs!

In 1930 (when Pluto was discovered) no disc could be discerned. If we take 0.4 arcsec as the maximum size of Pluto's disc, then at the distance from the Earth to Pluto in 1930 (Figure 2.12), Pluto's diameter could not exceed something like 11 000 km. This is roughly the same as the Earth's diameter (12 756 km), but 11 000 km is an upper limit. It could be argued that the 0.4 arcsec limit is a bit pessimistic, and perhaps a smaller limit is more realistic, particularly when the seeing is excellent. However, the Dutch-American astronomer Gerard Peter Kuiper (1905–1973) stated in 1949 that, using the 82-inch (2.08-metre) reflector at the McDonald Observatory in Texas, the upper limit of the angular diameter of Pluto was indeed 0.4 arcsec. In 1950, using the 200-inch (5.08-metre) reflector on Mount Palomar in California, Kuiper obtained the significantly lower value of 0.23 arcsec for the *actual* size of Pluto, which implies very steady atmospheric conditions. The latter value corresponded to a diameter of about 5800 km for Pluto. Fifteen years later, in 1965, astronomers at the US Naval Observatory (USNO) in Flagstaff could only obtain an upper limit of 0.25 arcsec, which is about the same as Kuiper's 1950 claimed actual size. The angular diameter of Pluto, even in 1965, was clearly shrouded in uncertainty.

Another approach to obtaining the size of Pluto is based on the amount of sunlight it reflects towards our telescopes. If we know the reflectivity of Pluto's surface then, from the amount of reflected radiation collected by a telescope, the area of Pluto's surface can be calculated and therefore its volume too. But how is the reflectivity of Pluto's surface obtained? The answer is by making reasonable assumptions about the composition and roughness of its surface. For example, a black, tarry surface reflects a lot less sunlight than a snowy surface. If a darker surface is assumed than is actually the case then the size of Pluto will be over-estimated, and vice-versa.

Early estimates using this method gave Pluto a diameter of a little over 6000 km. Other estimates were different, because of different assumptions about its surface, though the smallest values were only a few times smaller than the largest values. It was pretty clear that Pluto was not bigger than the Earth, and could be substantially smaller. Then, in 1976, Dale Cruikshank, Carl Pilcher and David Morrison of the University of Hawaii, established, from measurements of Pluto's reflectivity, that Pluto is a very good reflector of sunlight, approaching that of fresh snow. The amount of reflected sunlight could then be accounted for by a smaller surface than were the surface darker. It was at once clear that Pluto is far smaller than the Earth. Note that 'reflectivity' is a very loose term; the proper term is 'geometric albedo' (Box 3.1), which I'll abbreviate to 'albedo' and use from now on.

An accurate value at last

You will see in Section 3.2 that in 1978 Pluto was discovered to have a satellite, subsequently named Charon. Just three weeks after the discovery was announced, the Swede Leif Erland Andersson (1944–1979), at the Lunar and Planetary Laboratory in Tucson Arizona, made what turned out to be a crucial prediction in the quest to get an accurate value for Pluto's size. Charon orbits in Pluto's equatorial plane, which is inclined at $57.5°$ with respect to its orbital plane. Andersson realised that, as viewed from the Earth, there would come a time when Charon's orbit is presented edgewise to us. At and near such times Charon would pass between us and Pluto, and half of Charon's orbital period of 6.387 days later it would pass behind Pluto.

BOX 3.1 A BODY'S AREA FROM ITS REFLECTIVITY,
AND ITS VOLUME FROM ITS AREA (FOR THOSE
COMFORTABLE WITH ALGEBRA)

Area from reflectivity (albedo)
From the reflected radiation collected by a telescope a quantity
called the flux density can be calculated. Flux density, F, is a general term defined as the power of the electromagnetic radiation
(such as sunlight) incident on unit area of a receiving surface (such
as a telescope mirror or lens). Our receiving surface will be a telescope aperture perpendicular to the direction to the body (such as
Pluto).

Let's assume that the body is in opposition, so that the Sun,
Earth and the body are near enough in a straight line. In this case
the body is seen from the Earth at what is called zero phase angle,
as in Figure Box 3.1, and the flux density received by reflection is
labelled $F_r(0)$. It can be shown that

$$F_r(0) = kpA,$$

where k is a combination of known factors involving the Sun and
the distance to the body, A is the projected area of the body in our
direction (Figure Box 3.1 (a)) and p is a quantity called the albedo.
This is the ratio $F_r(0)/F_L(0)$, where $F_L(0)$ is the flux density we *would*
have received from a flat Lambertian surface perpendicular to the
direction to the Sun and Earth, and with an area equal to the projected area of the observed body (Figure Box 3.1 (b)). A Lambertian
surface is perfectly diffuse (the opposite of a mirror) and reflects
100% of the radiation incident upon it. I expect you can see that
it is reasonable that $F_r(0)$ increases as p increases and also as A
increases.

If we know p, and if we have measured $F_r(0)$, then A can
be obtained from the equation. For bodies with negligible atmospheres, the value of p depends on the composition and roughness

of the surface. If the surface is fairly smooth then, in opposition, it can scatter a comparatively large amount of sunlight towards the Earth; p can even exceed 1. This brightening around opposition is called the opposition effect. Slightly away from opposition p is less than 1. Clearly, in seeking candidate materials to match the observations, not only must composition and roughness be considered, but also the angle between our line of sight and the direction from the body to the Sun.

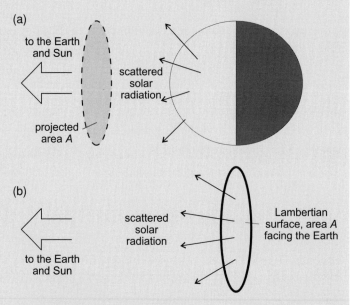

FIGURE BOX 3.1 (a) A spherical body (such as Pluto) in opposition. (b) A flat Lambertian surface with the same projected area as the body.

Well away from opposition, the area we see illuminated from the Earth is less than A, which also needs to be taken into account. The details will not concern us.

Volume from projected area
How is the volume, V, calculated from A?

Planets are near enough spherical, and therefore for a planet with a radius R the volume, V, is

$$V = \frac{4}{3}\pi R^3,$$

where π ('pie') is the circumference of a circle divided by its diameter. It is a curious number, one that goes on without end, 3.141592653 ... It is an example of a transcendental number. The area, A, of a circular disc with a radius R is given by $A = \pi R^2$ and therefore

$$R = \sqrt{\frac{A}{\pi}}.$$

V is then obtained from the previous equation.

The former is an example of a transit, the larger body being only partially obscured by the smaller body, and the latter an example of an occultation if the larger body completely obscures the smaller body. In both cases dips in the combined brightness of Pluto plus Charon will occur.

Andersson calculated when transits and occultations would next occur, using the best estimates of the sizes of Pluto and Charon, and realised it would be in the 1980s. What good fortune – it could have been much further into the future. These transits/occultations could be detected at that time by measuring the brightness of the Pluto-Charon system with a photomultiplier tube (PMT), or with an array of small electronic detectors called a charge-coupled device (CCD). Many astronomers at various observatories attempted to detect the transits. In February 1985 the first definite evidence of the event was obtained by the American astronomer Richard Binzel, using a PMT with the 0.9-metre reflector at the McDonald Observatory in Texas. He recorded a dip in the combined brightness of Pluto plus Charon. Exactly 6.387/2 days later the American astronomer David Tholen, using a PMT with the 2.24-metre reflector of the University of Hawaii on Mauna Kea, recorded another dip, exactly as expected.

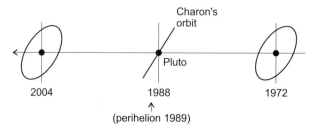

FIGURE 3.1 The circular orbit of Charon as seen from the Earth at three different dates. The orientation of Charon's orbit is fixed with respect to the stars. (Not to scale)

The orbital configurations, as we see them, are illustrated in Figure 3.1. Transits/occultations lasted from 1985 to 1991. Since then, as Pluto moves around its orbit Charon's orbit continues to widen. It will then get narrower until transits/occultations again occur. The whole sequence is repeated every half period of Pluto's eccentric orbit, 249/2 years, though next time the events will occur with Pluto near aphelion.

The size of Pluto was obtained from the maximum duration of the many events that occurred from 1985 to 1991. So-called central events started in March 1987, which is when the whole of Charon's disc passes between us and Pluto's disc, and when Pluto completely obscures Charon. Though Charon's orbit had earlier enabled Pluto's mass to be obtained (Section 3.2), uncertainties in Pluto's density led to calculations of its diameter ranging from 1900 to 4300 km. Now, by selecting the duration of the longest transit of the centre of Charon, the diameter of Pluto is given by the duration multiplied by the known orbital speed of Charon (0.220 km per second). This is not quite straightforward, for example from Earth a small proportion of the shadow of Charon on Pluto would affect the amount of reflected sunlight reaching the detector. But corrections for this can be made, and adjustments for other small complications were made too. The result was 2400 km for the diameter of Pluto.

The biggest source of uncertainty was the 5% value in the semi-major axis of Charon's orbit, because that affects the orbital speed of

FIGURE 3.2 The Hubble Space Telescope, launched in 1990. It has a main mirror 2.4 metres in diameter. (NASA) (See plate section for colour version.)

Charon. In the 1990s the Hubble Space Telescope, with its 2.4-metre diameter mirror, untroubled by atmospheric turbulence (Figure 3.2), reduced this uncertainty to 1.5%, and established the best value to be 19 570 km. The corresponding diameter of Pluto is 2360 km, with an uncertainty of ±60 km. Subsequent observations, including the use of ground-based telescopes fitted with optical devices to combat atmospheric turbulence (adaptive optics), has yielded a diameter of 2274 ± 16 km, quoted by the Jet Propulsion Laboratory (JPL).

On 12 June 2006 Pluto occulted a star. The previous four occultations occurred in 2002 (two), 1988 and 1985. From precise knowledge of the track behind Pluto of the occulted star a value of Pluto's diameter can be obtained. More than one observatory observes such rare events and, from the 2006 occultation, observed from five sites in Australia, a value of 2304 ± 64 km was obtained, slightly larger than the JPL value. The uncertainty has since been reduced to about ±20 km.

Note that during the stellar occultation the star faded slowly rather than in an instant. This indicates that Pluto has a thin atmosphere (Section 5.3). But it also makes Pluto's surface more difficult to detect; this accounts for much of the (small) uncertainty.

3.2 PLUTO'S MASS

In Section 1.6 I outlined how the mass of a body in space is obtained from its gravitational effect on another body, and examples were given – a satellite orbiting a planet, a spacecraft flying past it. It was not until 1978 that the first of Pluto's satellites was discovered, and no spacecraft or any other object has ever flown past Pluto. So how was Pluto's mass estimated before 1978?

Pluto's gravitational effect on Neptune

You saw in Section 2.4 that as soon as Pluto was discovered in 1930 it was realised from its unexpected faintness that it was very unlikely to be anywhere near the 6.6 M_E predicted by Lowell for his Planet X to account for the discrepant motion of Uranus. Could Pluto's mass be constrained based on the motions of Uranus and Neptune?

In 1930, with the orbit of Pluto known, estimates were made of Pluto's mass based on the imperceptibility of its gravitational effect on Uranus and Neptune, with values ranging from 0.1 M_E to 1 M_E. It was in 1931 that the American astronomer Seth Barnes Nicholson (1891–1963) attempted to use the orbit of Neptune to calculate the actual mass of Pluto. He obtained 0.94 M_E with an uncertainty either way of 25%. In the following 20 years or so a few other astronomers used Nicholson's approach and came up with values of a little over 0.9 M_E, but less than 1 M_E. In spite of some scepticism about these values, they stood as reasonable estimates for some years.

It was in 1968 that the next major revision to Pluto's mass was made, again based on Neptune's orbit. In that year a team of astronomers from the US Naval Observatory (R L Duncombe,

W J Klepczinski and P K Seidelmann) deduced that the mass of Pluto was only 0.18 M_E, a result arrived at independently a year later by the American astronomer Dennis Rawlins. The USNO team revised Pluto's mass downwards in 1970, to 0.17 M_E, and in 1971 to just 0.11 M_E. In that same year the American astronomers J G Williams and G S Benson came up with 0.17 M_E.

Turning sizes into masses

Another approach is to obtain Pluto's mass from its size. To turn a size into a mass the mean density (the whole-planet density) of Pluto has to be used. Pluto's mass is then Pluto's volume times Pluto's mean density. Densities of the sort of substances that could comprise Pluto include water ice (density around 1000 kg per cubic metre – 1000 kg/m^3), carbon-rich solids (1000–2300 kg/m^3), silicates (2600–3300 kg/m^3) and iron-rich substances (e.g. Fe-Ni, 7295 kg/m^3). These are the densities at the Earth's surface, but Pluto is so small that pressures in the interior would not raise the densities of these solids significantly. The range of densities provides a lot of freedom to mod-ellers, with plausible mean densities of Pluto varying from a little over 1000 kg/m^3 for a world dominated by ices (predominantly water ice but not excluding other icy materials) to about three times greater for a rocky world (perhaps with a small fraction of Fe-Ni).

Kuiper's 1949 upper limit of 0.4 arcsec for the angular diameter of Pluto (see previous subsection) led him to give an upper limit of the diameter of 10 300 km. His quoted upper limit on the mass, 0.8 M_E, corresponds to an unrealistically high mean density of 9200 kg/m^3 for Pluto. His 1950 claimed measurement of 0.23 arcsec gives Pluto a diameter of 5800 km. The quoted mass, 0.1 M_E, corresponds to a mean density of 4600 kg/m^3 for Pluto – somewhat lower than the value for the Earth, 5520 kg/m^3. The 1965 upper limit of 0.25 arcsec from the USNO, and the quoted upper limit on the mass of 0.14 M_E correspond to a mean density of 6000 kg/m^3.

The conclusion by Cruikshank *et al.* in 1976 that Pluto has a highly reflecting surface (see previous subsection), and was therefore

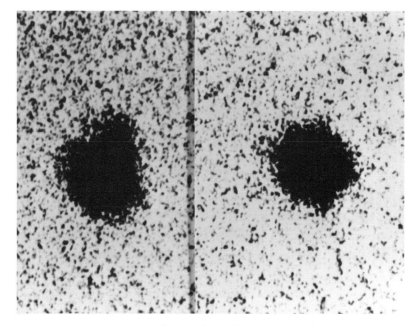

FIGURE 3.3 Images of Pluto obtained in 1978 by Anthony V Hewitt. The disc is not resolved. The elongation in the left image is Pluto's satellite Charon. In the right image its angular separation from Pluto is too small for it to be seen. (NASA, USNO)

smaller than if it had been darker, led them, via an *assumed* density, to a mass of only 0.003 M_E, about a quarter that of the Moon.

Oh dear, just how small *was* the mass of Pluto? The answer came in 1978.

A measured value of mass at last

Figure 3.3 shows an image of Pluto obtained in 1978 by the astronomer Anthony V Hewitt at USNO using the 1.55-metre reflecting telescope. He had obtained images on 13 April, 20 April and 12 May, with two photographs on each date and three images of Pluto on each photograph. The disc of Pluto is not resolved; the images in Figure 3.3 (and his other images) are spread out by atmospheric turbulence as described earlier. Because of the elongation the images had been

marked as 'poor', which could have been caused by a telescope track-
ing error during the 90 second exposures. On 22 June 1978 James W
Christy (USNO) requested images of Pluto to assist the USNO pro-
gramme to establish a more accurate orbit for the planet. He noticed
that though the image of Pluto was elongated, the stellar images were
not. Moreover, in the 13 and 20 April images the elongation was to
the south, whereas on 12 May it was to the north. The following day,
23 June, Christy obtained fifty earlier USNO images of Pluto. Two
images from 1965 showed the bulge, and a sequence of five taken in
one week in 1970 showed the elongation moving clockwise in about
six days. Christy concluded that Pluto had a satellite. The astronomer
Robert S Harrington (1942–1993), also at USNO, computed an orbit
assuming that the orbital period was 6.387 days, the same as Pluto's
rotation period (Section 4.5). His orbit fitted the positions of Pluto's
elongations on all the images very well.

The satellite was named Charon, the boatman in Greek mythology, who ferried the souls of the dead across the River Styx to Hades,
where Pluto ruled.

In Section 1.6 I stated that the mass of a planet can be obtained
from the orbit of a satellite of far lower mass than the planet. To
accomplish this the orbital period of the satellite and its semimajor
axis need to be known. Images such as that in Figure 3.3 indicate
that Charon's size, and therefore its mass, cannot be assumed to be
far lower than that of Pluto. In this case the orbital data provide the
combined mass of Charon and Pluto. The angular separation between
Charon and Pluto had been measured to be 0.8–0.9 arcsec, which
corresponds to a separation of about 17 000 km. In a near-enough
circular orbit this is the required semimajor axis of Charon's orbit
around Pluto. With a known orbital period of 6.837 days, Harrington
calculated the combined mass to be about 0.0017 M_E. In 1980 this was
revised to 0.0023 M_E. The faintness of Charon compared with Pluto
led Christy and Harrington to conclude that 90–95% of the mass of
Pluto-Charon was Pluto, which gave Pluto a mass of about 0.0022 M_E,
small indeed!

FIGURE 3.4 A Hubble Space Telescope image of Pluto and Charon taken in 1994. The discs are clearly resolved. (R Albrecht, ESA/ESO Space Telescope European Coordinating Facility, NASA)

You saw in Section 3.1 that the 12 June 2006 stellar occultation by Pluto resulted in Pluto's diameter being established as 2304 km. Charon's diameter was obtained earlier, from a rare stellar occultation in April 1980. From Charon's orbital speed and the 20 second duration of the occultation, Charon's diameter was determined to have a value of 1200 km, with only a small uncertainty. It is therefore about an eighth of Pluto's volume, and if it has the same mean density, Pluto would account for just under 90% of the Pluto-Charon mass, not significantly different from Christy and Harrington's lower limit.

In 2003 Hubble Space Telescope images like that in Figure 3.4 were used to obtain an accurate determination of the position of the centre of mass of Pluto plus Charon, which enabled the ratio of their masses to be calculated accurately, 8.2:1, with an uncertainty of only 6% (Box 3.2). From the value of 0.0023 M_E for the combined mass it then followed that Pluto's mass is 0.0022 M_E with an uncertainty of 5%.

The uncertainty was reduced to 0.4% in 2008 when David Tholen and colleagues obtained the mass of Pluto and, simultaneously, the masses of its *three* satellites, the other two being tiny, from

BOX 3.2 MASS RATIOS OF TWO BODIES WHEN THE POSITION OF THE CENTRE OF MASS IS KNOWN (FOR THOSE COMFORTABLE WITH ALGEBRA)

The centre of mass of two balls on Earth joined by a thin stick is at the balance point, which will clearly be nearer the more massive of the two. If M and m denote the masses of the more massive and less massive body respectively, and d and D are, respectively, the distances of M and m from the centre of mass, then (as you could confirm with a simple experiment)

$$M \times d = m \times D,$$

which can be re-arranged as

$$M/m = D/d.$$

So, with M greater than m, D, the distance of m from the centre of mass, is greater than d. If, for example M/m is 10, then D/d is also 10, i.e. m is 10 times further from the centre of mass than is M.

These equations also apply to two bodies orbiting each other in space, such as Pluto and Charon.

the orbits of all the bodies within the Pluto system (Section 4.2). The mass obtained for Pluto is 0.00218 M_E.

Pluto's actual mass is thus smaller than almost all the earlier estimates. Pluto is certainly not Lowell's Planet X, nor anyone else's pre-discovery predictions of one or more planets beyond Neptune. Recall from Section 2.4 that it was established in 1993 after Voyager 2 visited Neptune that there are no discrepant motions of Uranus and Neptune. The earlier beliefs that there were depended critically on the time spans of the orbital data used and on their accuracy.

3.3 PLUTO'S DENSITY AND GLOBAL COMPOSITION

With a mass of 0.00218 M_E and a diameter of 2304 ± 20 km, the mean density of Pluto is 2030 ± 60 kg/m^3. It is beyond reasonable doubt

Table 3.1 *Densities and melting temperatures of common solids for planet building.*

	Density[1] (kg/m³)	Melting temperature[2] (°C)	(K)
Rocky materials and iron			
Carbon rich (e.g. tars)	1000–2300	A few hundred	°C values + 273
Common Earth rocks	2600–3300	Around 1000	°C values + 273
Iron (with 6% nickel)	7925	1492	°C values + 273
Icy substances[3]			
Water, H_2O	996	0	273
Carbon dioxide, CO_2	1920	−56.6	216.4
Ammonia, NH_3	820	−77.7	195.3
Methane, CH_4	494	−182.5	90.5
Carbon monoxide, CO	1240	−205	68
Nitrogen, N_2	1490	−210	63

[1] Densities increase with pressure, but these values are indicative of the density of the material in Pluto's interior.

[2] K stands for Kelvin, the temperature scale that starts from absolute zero, at minus 273°C.

[3] For icy substances the triple point temperatures are given for the melting temperatures.

that Pluto consists of a mixture of substances. Recall from Section 3.2 that likely substances include icy materials, carbon-rich solids, silicates and iron-rich compounds. Among the icy substances water will be the most abundant, though smaller quantities of other icy substances are to be expected, and indeed these are found on Pluto's surface (Section 5.2).

Table 3.1 lists the solid densities of these and other substances. The values increase slightly as pressure increases, but can be taken as indicative of values in Pluto's interior. The temperatures are those at which the substance melts. The values are dependent on pressure. For rocky materials, Fe-Ni and water, the melting temperatures are at Earth's atmospheric pressure, but the pressure dependence is very slight. For the icy materials the pressure dependence is somewhat greater: the temperatures in Table 3.1 are at what is called the triple point pressure, and the temperature is called the triple point temperature. The triple point is where the gaseous phase, the liquid phase and the solid phase can coexist in equilibrium (more on this in Section 5.3). The important point here is that the triple point temperature shows in the correct order the 'freezability' of the various ices, with water remaining solid to higher temperatures than the other icy materials. Pluto's small size and great distance from the Sun mean that its temperatures throughout are likely to be so low that many or all of its icy materials will be frozen solid.

You can well appreciate that, apart from excluding a significant proportion of Fe-Ni, a global mean density of around 2000 kg/m^3 can be obtained by a range of compositions. If, as is likely, icy and rocky materials predominate, then about a third of the mass will be icy materials, particularly water, the rest consisting of partially hydrated rocky materials and perhaps some Fe-Ni. The overall reflectivity (more precisely the albedo) of Pluto is 55%, indicating that much of the surface is as reflective as freshly fallen snow on Earth. The common icy materials all form bright white deposits at Pluto's low average surface temperatures of about − 230°C, which is only about 43°C above absolute zero, i.e. about 43 K (K stands for Kelvin, the temperature scale that starts from absolute zero at − 273°C).

There is also the question of whether the different substances are mixed throughout the interior of Pluto, or whether there is layering, and if so to what extent. These questions will be addressed in Section 5.4.

The relative abundances of the chemical elements
On what evidence are my above phrases 'likely substances include icy materials, carbon-rich solids, silicates and iron-rich compounds',

and 'If, as is likely, icy and rocky materials predominate . . . . . . particularly water' based? The phrases spring from the relative abundances of the chemical elements in the Solar System, established from a great range of measurements using many different techniques. Table 3.2 shows the outcome for the 15 most abundant chemical elements.

In considering the composition of the non-giant planets, hydrogen and helium can be discounted, because even at Pluto's low temperatures these elements, including the molecule H_2, are gases and cannot be retained by a small body like Pluto. But for many compounds of hydrogen this is not so. The most abundant compounds of hydrogen, as you can infer from Table 3.2, are CH_4 (methane), NH_3 (ammonia) and H_2O (water), with water the most abundant. To these icy materials must be added CO (carbon monoxide) and CO_2 (carbon dioxide), the relatively high abundances of which are also consistent with Table 3.2.

Rocky materials are dominated by silicates, compounds of silicon, oxygen and one or more metallic element, notably magnesium and iron, all relatively abundant. Iron itself is next to magnesium in metallic abundances, and almost always comes alloyed with nickel (Fe-Ni).

Carbon-rich solids contain compounds made of carbon and hydrogen, and perhaps other elements. Substances free from carbon, such as silicate or Fe-Ni grains can be present as minor components in carbon-rich solids. CH_4 (in solid form) is an example of a carbon rich solid that you've already met. This is a small molecule; other carbon rich molecules can contain far more atoms.

If chemistry is new to you, you might wonder why neon has not been mentioned, nor compounds of helium and neon; they are both abundant in the Solar System. This is because (like argon) they are chemically inert – they almost never form compounds – and are called inert gases. They remain gases to extremely low temperatures, down to 4.2 K for helium and 27 K for neon (both values at low pressures).

Table 3.2 *Relative abundances of the 15 most abundant chemical elements in the Solar System.*

Chemical element			Relative atomic mass[2]	Relative abundance[3]	
Atomic number[1]	Name	symbol		by number of atoms	by mass
1	hydrogen	H	1.0080	1 000 000	1 000 000
2	helium	He	4.0026	97 700	388 000
6	carbon	C	12.0111	331	3 950
7	nitrogen	N	14.0067	83.2	1 160
8	oxygen	O	15.9994	676	10 730
10	neon	Ne	20.179	120	2 410
11	sodium	Na	22.9898	2.09	48
12	magnesium	Mg	24.305	38.0	917
13	aluminium	Al	26.9815	3.09	83
14	silicon	Si	28.086	36.3	1 010
16	sulphur	S	32.06	15.9	504
18	argon	Ar	39.948	2.51	100
20	calcium	Ca	40.08	2.24	89
26	iron	Fe	55.847	31.6	1 750
28	nickel	Ni	58.71	1.78	104

[1] The atomic number is the number of protons in the nucleus. This is unique to each element. Elements with atomic numbers from 1 to over 100 are known, with no gaps.

[2] The relative atomic mass is with respect to the isotope of carbon that has, in addition to six protons, six neutrons, so is known as carbon-12, or ^{12}C. For all the elements, the non-integer values in the table are averages over their naturally occurring isotopes.

[3] Abundances are given to three significant figures. Many values are known to greater accuracy. The helium values correspond to those before the conversion of some of the hydrogen in the Sun's core to helium, i.e. to the Sun at its formation.

The inert gases would be expected to be present on Pluto as no more than traces of gas.

I hope you can now see why specific substances have been picked out as candidate materials for building Pluto.

In the next chapter I'll include Pluto's family of three satellites.

4   Pluto's family

As well as Charon, Pluto has two other satellites, both very tiny. Pluto is thus the largest member of a small family. In this chapter the emphasis is on the family as a whole.

The two tiny satellites were discovered in May 2005 by H A Weaver and colleagues, with the Hubble Space Telescope during studies of Pluto in relation to NASA's *New Horizons* spacecraft, now on its way to Pluto (Chapter 8). The discovery team proposed the Greek names Nyx and Hydra, but as Nyx already appeared in the names of two asteroids, the spelling was changed from the Greek Nyx to the Egyptian Nix. In mythology Nix (Nyx) is the goddess of darkness and light, very appropriate for a satellite orbiting the god of the underworld. Nix is also the mother of Charon – rather a small mother for such a large child! Hydra is a nine-headed serpent, a suitable name for the ninth planet Pluto. In June 2006 the International Astronomical Union approved these names.

Figure 4.1 shows an HST image of Pluto and all three satellites obtained on 15 February 2006. Pluto and Charon are grossly overexposed so that the very faint Nix and Hydra can be imaged. Nix and Hydra are the two dots, Hydra being the further from Pluto.

4.1 REVOLUTIONS AND ROTATIONS

Revolutions (orbits)
Figure 4.2 shows the orbits of the three satellites, to scale, but viewed face-on, a view denied to us from the Earth by the orbital inclinations. Beneath the orbits, Pluto and its three satellites are shown to scale. Table 4.1 lists properties of the orbits.

FIGURE 4.1 Pluto and its three satellites, imaged by the Hubble Space Telescope on 15 February 2006. Pluto (overexposed) is at the centre. Charon (also overexposed) is nearest Pluto. Nix and Hydra are the two dots, Hydra being the further from Pluto. (NASA, ESA, H Weaver, A Stern and the HST Pluto Companion Research Team)

All three orbits are very nearly in the same plane (Hydra is only 0.2° off). The orbits lie in the equatorial plane of Pluto. This plane makes an angle of 57.5° with respect to the plane of Pluto's orbit around the Sun. The satellite orbits are all of low or zero eccentricity. This is to be expected for orbits that are rather small in comparison with Pluto's diameter of 2304 km; if the eccentricities were initially higher, they would be reduced by gravitational interactions with Pluto. The orbits are retrograde, the opposite to the prograde direction of most of the satellite orbits, and all planetary orbits in the Solar System (Section 1.2).

Table 4.1 *The orbits of Pluto's satellites.*

	Semimajor axis (km)	Orbital period (days)	Eccentricity	Orbital inclination (°)
Charon	19 570	6.387	0.00	0.0
Nix	48 680	24.856	0.02	0.0
Hydra	64 780	38.207	0.05	0.2

The orbital inclinations are with respect to the equatorial plane of Pluto. Data from The Royal Astronomical Society of Canada's *Observer's Handbook 2009*.

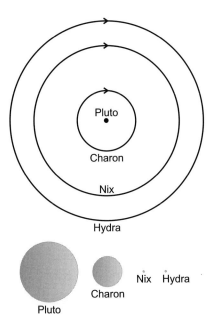

FIGURE 4.2 The orbits of Pluto's three satellites with respect to Pluto. The orbits are in the same plane, the plane of Pluto's equator, and are circular. They are shown face-on, but the inclination of their orbital plane means that we don't get this face-on view from the Earth. Beneath the orbits, Pluto and its three satellites are shown to scale.

The orbital periods are in the ratios 1:4:6 (Charon:Nix:Hydra). With Charon so much more massive than Nix and Hydra, gravitational interactions of Nix and Hydra with Charon have resulted in these simple ratios, which, once acquired, are stable. These are further examples of mean motion resonance (Section 2.5).

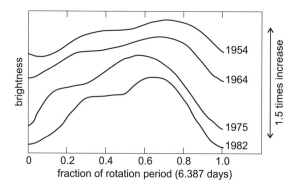

FIGURE 4.3 Lightcurves for Pluto obtained in the years shown. Each curve in each year repeats itself every 6.387 days, which must be the rotation period of Pluto. Each curve shown is typical for its year. (Adapted from S A Stern *et al.*, *Icarus* (1988))

Rotation – Pluto

There are two aspects of Pluto's rotation to be considered: the orientation of its rotation axis, and its axial rotation period. (The latter is with respect to the distant stars, though Pluto moves so slowly around its orbit that this is very nearly the same as its rotation period with respect to the Sun.)

It was in 1955 that the first good determination of Pluto's rotation period was made, from Pluto's lightcurves. A lightcurve is a graph of the brightness of a body versus time. For a spherical body with no surface features the light 'curve' is a horizontal straight line – there is no brightness variation. The 1955 light curves for Pluto were obtained by Robert Hardie and Merle Walker of Vanderbilt University, using a 1.1-metre reflector fitted with a PMT, a device that had been available to astronomers for only a few years. The lightcurve showed a brightness range of about 10%. This indicated surface features. Atmospheric features were ruled out by the fact that the lightcurve repeated every 6.39 days, with an uncertainty of only 4 minutes (0.003 days). Figure 4.3 shows lightcurves for Pluto obtained in four well separated years. The repeat time is the same in each year, which must be the rotation period of Pluto; atmospheric features would not display such constancy. Subsequently the value was refined to 6.387 days.

Where have you recently met 6.387 days? This is the orbital period of Charon. So, its orbital period is the same as the rotation period of Pluto. It is also in the same direction. This has resulted from the gravitational interaction between Pluto and Charon. More specifically it has arisen from the tide raised by Charon in the whole body of Pluto. This process is outlined in Box 4.1, should you wish to learn further details.

Analysis of lightcurves was also the first technique used to establish the orientation of Pluto's rotation axis. In 1972 Leif Andersson, during his PhD research at Indiana University in Bloomington USA, and John Fix of the University of Iowa, showed that lightcurves of the sort in Figure 4.3, where the broad features remained the same from year to year, could be used to estimate the axial inclination of Pluto, i.e. the angle that Pluto's rotation axis makes with respect to its orbital plane (an angle also called the obliquity). That the broad features remain the same whereas the mean brightness and its range vary, can be explained by the change in the range of latitudes presented to the Earth as Pluto moves around its orbit; it does not require changes in Pluto's surface features. Using a computer model they concluded that the axial inclination of Pluto was at least 50° and perhaps significantly larger.

Further clues came from the orbit of Charon (discovered in 1978). The orbit was seen to be inclined at 58° with respect to the orbital plane of Pluto. For dynamical reasons it was expected that the orbit of Charon would lie in the plane of Pluto's equator, regardless of whether Charon formed along with Pluto or was captured. Therefore, the axial inclination of Pluto was likely to be 58°. Subsequent images of Pluto showing surface features confirm this to be pretty well the case, and that the axial inclination of Pluto is 57.5°. Note that Pluto's axial rotation is retrograde, like its satellite orbits. Note also that calculations have shown that Pluto's axial inclination varies with a period of 3 Myr, from about 54° to about 78°, and back again; the satellite orbits will always be in Pluto's equatorial plane.

BOX 4.1 TIDAL INTERACTIONS (FOR THOSE WISHING TO GO DEEPER)

You are doubtless familiar with the tides raised in Earth's oceans by the Moon and the Sun. In fact the Moon and the Sun raise tides in the whole body of the Earth, though these are smaller than the ocean tides because of the strength of the Earth's solid mantle and crust. In this Box my concern is with whole body tides.

Figure Box 4.1(a) shows bodies A and B falling freely towards each other because of their mutual gravitational attraction. The distortion induced in the initially spherical body A by the gravity of body B, and the distortion induced in initially spherical body B by the gravity of body A are each called tidal distortions, or tides. These distortions arise because the gravitational attraction between a small bit of one body and a small bit of the other body decreases as the distance between the two bits increases, the lighter the shade in Figure Box 4.1(a) the greater the gravitational force of attraction. When allowance is made for the internal gravity and internal strength of each body, both of which tend to keep it spherical, the outcome is a shape a bit like a rugby ball (or an American football).

Figure Box 4.1(b) shows the orbit of B around A with the rotation axes of A and B perpendicular to the plane of the orbit (this does not need to be exactly so). A and B are still falling towards each other but their sideways motion results in orbital motion. A is rotating with a rotation period shorter than the orbital period of body B. As you can see, this causes the tidal distortion of A to be carried ahead of the line joining the two bodies, and it could only keep in line if the tidal distortion were to move 'backwards' in the direction opposite to the rotation of A. But the internal strength of body A is preventing such alignment. However, the parts of body A closer to body B experience a greater gravitational force than those parts of A further from B, which results in a twisting force on A, called a torque. This reduces the rotational period of A, and if

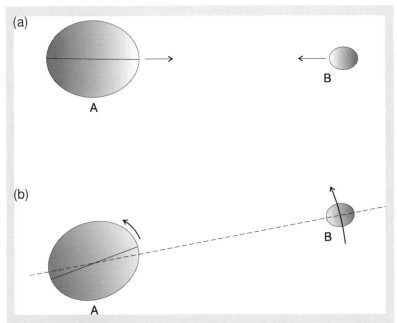

FIGURE BOX 4.1 (a) The tidal distortion in bodies A and B falling towards each other (greatly exaggerated), the lighter the shade the greater the gravitational force of attraction. (b) The orbit of B around A, where the rotation of A has not yet been slowed by B to equal the orbital period, whereas B has been fully slowed. The rotation of A carries its tidal bulge ahead of alignment with B.)

the torque is sufficiently large then, over a time considerably less than the age of the Solar System, the rotational period of A will be reduced to the orbital period of B around A, which is a stable configuration. In Figure Box 4.1(b) body B's rotation has already been fully slowed in this way.

Pluto (body A) and Charon (body B), have both had their rotations fully slowed which is why they always face each other.

There is a robust principle in physics called the principle of the conservation of angular momentum. In the present context, this means that if, as must have been the case with Pluto and Charon, their axial rotation periods have lengthened, then their rotational angular momenta must have decreased. With nowhere else to go this means that the angular momenta of the orbits has increased, in accord with the principle. In this case the size of the orbit will have increased whilst the present configuration was being achieved. Charon and Pluto are thus further apart than in the past, but they will move no further apart in the future. The details of the interactions that have resulted in this transfer of angular momentum will not concern us.

You might have noticed that the discussion above explains why the Moon keeps the same face towards the Earth. But why doesn't the Earth keep the same face towards the Moon? The answer is that whereas Pluto is 8.6 times the mass of Charon, Earth is 81.3 times the mass of the Moon. The Moon has certainly slowed down the Earth's rotation, but our satellite is not massive enough to have yet slowed it sufficiently. Another factor might be the greater distance to the Moon as a multiple of the Earth's diameter than is the case for Pluto and Charon; the present ratios are 30.1 and 8.5 respectively, though there's uncertainty about these ratios is the distant past.

Rotations – the satellites

Charon reflects only about 60% as much sunlight as Pluto, and whereas Pluto currently has a lightcurve amplitude of 38%, Charon's is only 8%. Nevertheless, this has been sufficient to establish a rotation period of 6.3972 days, rotating in the same direction as its orbit (retrograde). Also the rotation axis is surely perpendicular to its orbital plane. This means that Pluto and Charon always present the same faces to each other i.e. Charon is always in the same position in Pluto's Charon-facing sky, and Pluto is always in the same position in Charon's Pluto-facing sky. This is to be expected for two bodies relatively close together, as a result of tidal interactions between them (Box 4.1). Such interactions slowed down the rotations of Pluto and Charon until the stable face-to-face configuration was achieved.

Table 4.2 *Masses, diameters and mean densities of Pluto and its three satellites.*

	Mass (M_E)	Uncertainty in mass	Diameter (km)	Mean density (kg/m³)
Pluto	0.00218	0.5%	2304 ± 20	2030 ± 60
Charon	0.000254	3%	1212 ± 3	1630 ± 70
Nix	~1 × 10⁻⁷	90% !	~88	1630 assumed
Hydra	~0.5 × 10⁻⁷	200% !!	~72	1630 assumed

The mass of the Earth is 5.9742×10^{24} kg.

Of Nix and Hydra there is nothing to say. The lightcurves of these tiny bodies have so far been unobservable in our telescopes, so we know nothing about their rotations.

4.2 MASSES, SIZES, DENSITIES AND GLOBAL COMPOSITIONS

Masses and sizes

The masses of Pluto and its three satellites were obtained simultaneously in one fell swoop, by David Tholen and colleagues. The mass of Pluto was obtained from its satellite orbits in the same general way as when only Charon was known (Section 3.2), and the satellite masses were obtained from the slight modifications in their orbits due to the gravitational influences within the system. The results, published in 2008, are presented in Table 4.2, the uncertainties in mass being either way, so that, for example, the tabulated mass of Charon could be 3% too low or 3% too high. Note that the (highly uncertain) diameters of Nix and Hydra are not measurements, they are far too small to show discs, but are calculated by assuming mean densities the same as Charon's.

The diameter of Pluto was established as described in Section 3.1. The diameter of Charon was poorly known until a very rare event

---

BOX 4.2 OBSERVATIONS OF A STELLAR
OCCULTATION BY CHARON (OPTIONAL)
Three high altitude observatories, all in Chile, successfully
recorded Charon's stellar occultation. The instruments were: the
0.84-metre telescope at Cerro Armazonis, Antofagasta; the 2.5-
metre duPont telescope and 6.5-metre Clay telescope at Las Cam-
panas Observatory, La Serena; and the 8.0-metre Gemini South
telescope at Cerro Panchon, La Serena. The teams of astronomers
were the Paris Observatory Group led by Bernard Sicardy (Paris
Observatory), the MIT-Williams College Group led by Amanda
Gulbis (MIT) and the Southwest Research Institute-Wellesley
College group led by Eliot Young (SwRI). M J Person (MIT) and
colleagues published the results.

---

occurred on 11 July 2005 – a stellar occultation. It was observed
from several observatories (Box 4.2), and from the known track of
the star behind Charon's disc, the outcome is a mean diameter of
1212 ± 3 km, and a corresponding mean density of 1630 ± 70 kg/m^3,
substantially less than Pluto, indicating a greater proportion of icy
materials.

It was also discovered that Charon, like the Earth, is slightly
flattened at its poles. Remarkably, the different tracks of the star
behind Charon's disc revealed a depression, perhaps as much as 7 km
deep, at the edge of the disc of Charon.

You have probably noticed from Table 4.2 that the diameter
of Charon is better known than that of Pluto. This is because, as
mentioned in Section 3.1, the atmosphere of Pluto makes its surface
difficult to locate. By contrast, Charon has no discernible atmosphere.

Pluto and Charon are a remarkable pair. Their mass ratio of
8.6 is far and away the smallest ratio for any satellite system in
the Solar System. The Earth and Moon come next, with a ratio of
81.3. Pluto and Charon are also close in the sense that Charon's
orbital semimajor axis is only 8.5 times Pluto's diameter; for the

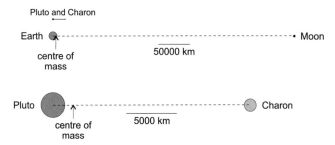

FIGURE 4.4 The Earth-Moon and Pluto-Charon systems. In the upper part of the figure both systems are shown to the same scale. In the lower part the Pluto-Charon system has been shown on a considerably larger scale. Note the positions of the centres of mass.

Earth and Moon the value is 30.1. Figure 4.4 compares these two systems.

Many astronomers regard Pluto and Charon as a double planet, though strictly Charon is too small to be regarded as a planet. An important criterion for deciding whether to call a system of two planetary bodies orbiting each other a double planet or a planet-plus-satellite is based on the centre of mass of the system. The centre of mass of two balls on Earth joined by a thin stick would be at the balance point. You can think of the centre of mass of a planet-satellite system in the same way. (If you read Box 3.2 you'll already know this.) The presence of the satellite means that the centre of mass cannot be at the centre of the planet. In the case of the Earth and Moon the centre of mass is therefore away from the Earth's centre, but is still inside the Earth, as shown in Figure 4.4. For Pluto and Charon Figure 4.4 shows that the centre of mass lies outside Pluto, though 8.6 times nearer to Pluto than to Charon (a factor necessarily the same as the ratio of their masses). It is this fact that has led to Pluto and Charon being regarded as a double planet.

It is now known that several Kuiper belt objects beyond Pluto comprise two fairly substantial bodies in orbit around each other with the centre of mass outside both of them, so Pluto-Charon is no longer unique.

I'll return to Pluto-Charon the double planet when we consider the origin of Pluto and its family in Section 4.3, and when the atmospheres, surfaces and interiors of Pluto and Charon are discussed in Chapter 5.

Densities and global compositions

The density and global composition of Pluto were the subject of Section 3.3; more on this in Section 5.4 where interior models of Pluto and its satellites are described. Here I'll focus on the density and global compositions of the three satellites.

The mean density of Charon, derived from its measured mass and diameter, is 1630 kg/m^3 (Table 4.2). The mean density of Pluto, obtained in the same way, is 2030 kg/m^3. This higher mean density of Pluto cannot be explained by greater compression in the interior of Pluto than in Charon: both bodies are sufficiently small that internal pressures are too low to raise the densities of the component materials much. Instead, it indicates that materials with lower intrinsic densities make up a greater proportion of Charon than they do of Pluto. In Section 3.3 it was indicated that icy materials, particularly water, comprise the low density component of Pluto, and the same is thought to be true for Charon. Icy materials (particularly water) are more abundant in Charon than in Pluto. In both cases the remaining mass is made up largely of rocky materials, partially hydrated, perhaps with some Fe-Ni.

It was noted in Section 3.3 that the high albedo (reflectivity) of Pluto, 55%, is consistent with very clean icy material covering the surface, like freshly fallen snow. Charon's albedo is only 35%, indicating that the icy material at the surface is either not very fresh or that it is mixed with darker materials. I'll return to the surface composition of Pluto and Charon in Sections 5.1 and 5.2.

Only the masses of Nix and Hydra have been measured, very imprecisely as Table 4.2 shows; there are no direct measurements of their sizes. If they are assumed to have the same mean density as Charon then the (very uncertain) sizes in Table 4.2 are obtained. From these sizes, and the amount of solar radiation that they reflect towards us, the albedos are 8% and 18% for Nix and Hydra respectively. You'll appreciate that these values are highly uncertain, but were they to be accurate then the surfaces of Nix and Hydra would be as dark as the Moon and Mercury, which, like Nix and Hydra, lack appreciable atmospheres, and darker than the other seven planets: very much darker than cloud-shrouded Venus, which has a albedo of 84%.

4.3 THE ORIGIN OF PLUTO AND ITS SATELLITES

In Section 1.5 I outlined the formation of the Solar System. Here we'll look more closely at the formation of Pluto and its satellites.

Pluto

You read in Section 1.5 that Pluto is one of many small objects in the outer Solar System that formed from icy and rocky materials that failed to grow into larger bodies, probably because of the shortage of materials in its feeding zone and gravitational stirring by the giant planets.

In the 1930s, and for several decades afterwards, another formation theory held sway. This was that Pluto was a satellite of Neptune that had been ejected. This was one of several theories put forward in the 1930s by Raymond Arthur Littleton (1911–1995). In the theory, Pluto is ejected through a close encounter with Neptune's most massive satellite, Triton (which had been discovered by William Lassell on 10 October 1846, just a few weeks after the discovery of Neptune itself, Section 2.2). In the 1930s Triton was the only satellite Neptune was known to have. Table 4.3 lists the orbital and physical properties of Triton, along with two much smaller satellites discovered later, including their dates of discovery.

The Table shows that Triton has an orbital inclination of greater than 90°, which means that it is in retrograde motion around its orbit (it could alternately be stated that the orbital inclination is $(180° - 159°) = 21°$ retrograde). The ejection theory, it was claimed, could explain this retrograde motion. It also explained Pluto's small size, having been a satellite comparable in size to Triton. Further

Table 4.3 *The three largest of Neptune's thirteen known satellites.*

	Semi-major axis [km]	Orbital period (days)	Eccentricity	Orbital inclination (°)	Mass (M_E)	Diameter (km)	Mean density (kg/m³)	Discovery
Proteus	117 600	1.122	~0	0.039	?	420	?	Voyager 2, 1989
Triton	354 800	5.877	~0	157	0.00358	2706	2060	W Lassell, 1846
Nereid	5 513 400	360	0.75	7.23	?	340	?	G Kuiper, 1949

The mass of the Earth is 5.9742×10^{24} kg.

Ten other satellites are known, the largest of which, Larissa, has a diameter of 194 km. All were discovered in or after 1989.

Data from The Royal Astronomical Society of Canada's *Observer's Handbook 2009*.

likely outcomes of ejection were claimed to be the high inclination and large eccentricity of Pluto's orbit and the crossing of Neptune's orbit by Pluto.

Wait a minute. You saw in Section 2.5 that the orbits of Pluto and Neptune do *not* cross and furthermore that the two planets never come close to each other. This would have been known in the 1930s, but I suppose it could have been argued that this is the situation today, and that close approaches billions of years ago could have occurred, and this must have been the case if Pluto was once a satellite of Neptune! It was only in the 1970s, with the growing computational speed of computers, that long term simulations of the orbits showed that the present avoidance *does* stretch back for billenia.

Though this long term orbital stability is sufficient to rule out the ejection theory, to this fatal blow was added William McKinnon's work in the 1980s, which showed that Pluto is insufficiently massive to have reversed Triton's orbit, but too massive to have been ejected by Triton.

When the ejection theory was shown to fail we had Pluto as

- an escaped asteroid? (No – dynamically unworkable.)
- a giant planet core that had missed out on gas capture because of slow growth? (No! – Pluto is far to small to have ever been a potential core, and its peculiar orbit is also unexplained.)
- a giant planet that lost its atmosphere? (No! – no mechanism is known that could do this to Pluto but, at the same time, not to Neptune.)

Triton: an aside

Before I turn to the presently accepted theory of Pluto's origin, I'll tie up the loose end of Triton's retrograde motion. The present view is that Triton was captured during the outward migration of Neptune (Section 1.5). Its retrograde orbit is a plausible outcome, depending on the details of the capture process. In any case, to turn flyby into capture Triton needed to lose orbital speed. This could have been the result of a skim through Neptune's atmosphere, or collision with, and destruction of, a smallish satellite; the ten not included in Table 4.3 range from Larissa (194 km diameter) to several just a few tens of kilometres in diameter, which indicates that many smallish satellites would have been present. Even so, if Triton-sized objects were rare in the space beyond Neptune, then capture would be highly improbable. However, the nebular theory of the origin of the Solar System (Section 1.5) leads us to believe that, not far beyond Neptune, there were many hundreds of objects with diameters exceeding 2000 km. One of these, Pluto, was and is in a stable orbit; one, we believe, was captured by Neptune to become Triton; the other large objects were in unstable orbits and so have been removed elsewhere.

Pluto and Triton are remarkably alike in size and density: Pluto 2304 km and 2030 kg/m^3, Triton 2706 km and 2060 kg/m^3. Also, they are in the same general region of the Solar System, which is consistent with the belief that there were many objects like them not far beyond Neptune. Interestingly, Triton has geysers, probably of nitrogen gas. Could Pluto have such geysers too? These would be beyond present imaging, but not beyond *New Horizons*, when it reaches Pluto in 2015 (Chapter 8). Figure 4.5 shows a *Voyager 2* image of Triton.

It is possible that the hugely eccentric orbit of the not insubstantial 340-km diameter satellite of Neptune, Nereid (Table 4.3), is a result of the capture.

Pluto

Here is the widely accepted theory of the origin of Pluto, which is inextricably linked to the theory of the origin of Charon. Pluto's origin was outlined in Section 1.5.

The low mean density of Pluto, which indicates a substantial fraction of icy materials, shows that it must have formed in the cold conditions of the outer Solar System, presumably beyond Neptune. It is thought to have remained small probably because of a shortage of material in its 'feeding zone', a result of the accretion timescale being so long that much of the material was lost through the gravitational influence of the giant planets and by the likely T Tauri phase of the

FIGURE 4.5 Neptune's largest satellite, Triton, which Pluto resembles, from the *Voyger 2* flyby in 1989. (NASA/JPL PIA02234)

young Sun. The increase in collisional speeds due to the gravity of the giant planets was surely another factor.

What about the large eccentricity and inclination of Pluto's orbit? The best current theory is linked to events on a larger scale, namely the outward migration of Uranus and Neptune during their formation 4600 Myr ago (Section 1.5). This swept the material beyond Neptune outwards, some of it into unstable orbits, some into far flung stable orbits, but some becoming trapped in the stable 3:2 mean motion resonance (Section 2.5). One of these was Pluto. Pluto and the many smaller bodies trapped in this resonance (Section 6.3) display a wide range of orbital eccentricities, which is in accord with models of the outcomes of a migrating Neptune. The values for Pluto's eccentricity and inclination are consistent with the modelling.

Charon

The favoured theory for the origin of Charon is that an object with a mass 30–50% that of Pluto collided with proto-Pluto, stripping away material from its outer regions into orbit, material from which Charon subsequently formed. This accounts for the lower mean density of Charon, provided that proto-Pluto had separated into a rocky-rich core and an icy-rich mantle, which is likely to have been the case (see Section 5.4).

Computer modelling shows that this mechanism could work. However, such a fruitful collision requires just the right impact speed and a strike on Pluto just the right distance from its centre. This is very unlikely unless, early in Solar System history, there were a lot more bodies in Pluto's region than there are today. (This is also required for the capture theory of Triton.) The existence of the Edgeworth-Kuiper belt beyond Neptune (Chapter 6) provides evidence that there were indeed a lot more bodies present.

Impetus for this collision theory doubtless came from work from the mid 1970s onwards indicating that a giant impact was far and away the best explanation of the origin of the Earth's Moon. Earlier theories comprised capture of the Moon by the Earth, fission of the Moon from a rapidly spinning Earth and accretion from debris in orbits around the Earth, left over after the formation of the Earth. None of these work dynamically nor do they explain why the Moon's composition is like a devolatilized portion of the Earth's mantle, which formed after segregation of its Fe-Ni core. Detailed computer simulations show that the impactor was an embryo roughly the mass of Mars (about 10% that of the Earth) and struck towards the end of planetary formation in the inner Solar System. Sufficient embryos would have been present to make it not improbable that one of the four terrestrial planets suffered such an impact.

Nix and Hydra

Could these tiny satellites be a by-product of the impact that formed Charon? No they couldn't; the debris disc from which Charon formed would not have extended to their orbits nor, according to computer simulations, would gravitational interactions with Charon move them out.

Could they be captured fragments of a collision between two bodies beyond Pluto and Charon? Again, no. In 2008 Yoram Lithwick and Yanqin Wu of the University of Toronto showed by means of numerical models that this is dynamically unworkable. They proposed instead that Nix and Hydra are captured bits of debris, planetesimals left over after the formation of the Solar System. This proposal remains plausible.

Let's turn now to the surfaces and atmospheres of Pluto and Charon, which will lead us on to consider internal models of these two bodies.

5   Surfaces, atmospheres and interiors of Pluto and Charon

So far, I've said very little about the surfaces and atmospheres of Pluto and its satellites, and for interiors I've given only the mean global densities and a broad indication of global compositions. In this chapter the compositions will be discussed in more detail, and internal models will be introduced.

I start with surfaces and atmospheres, which are clearly of intrinsic interest, but also because they provide clues and constraints about the interiors of Pluto and Charon. Very little is known about Nix and Hydra, therefore almost nothing is said here about these tiny satellites.

In Chapter 6 you will see that what we learn about Pluto and its satellites helps us to learn about other objects in the outer Solar System, notably the Kuiper belt objects.

First, some basic science. How do we obtain information about the surfaces and atmospheres of distant bodies? The answer is through measuring their albedos (reflectivities), which has already been discussed in Section 3.1, and by measuring their electromagnetic reflection and emission *spectra*.

5.1 REFLECTION AND EMISSION SPECTRA

The albedo of a body gives us information averaged over a wide range of wavelengths in solar radiation, particularly visible wavelengths. The reflectance spectrum is the reflectivity versus wavelength, and provides much more detailed information about a body. A reflectance spectrum is obtained with a device called a spectrometer, the essential components of which are illustrated in a simple way in Figure 5.1. The components are:

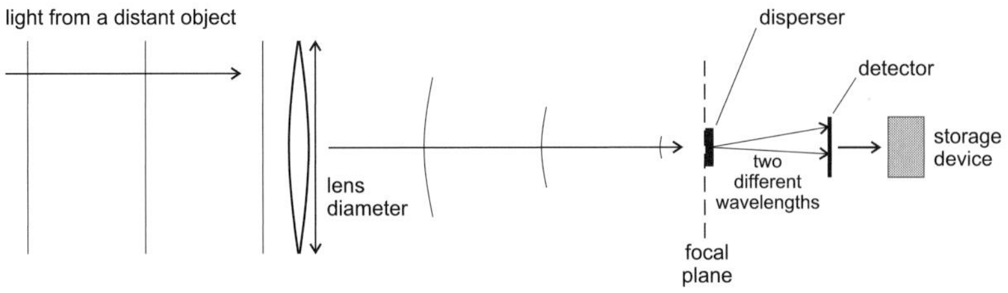

FIGURE 5.1 A simple diagram of the essential components of a spectrometer.

- a source of light (such as the Sun) to illuminate the distant object (the body being investigated)
- a lens or mirror (lens shown), to capture some of the reflected radiation
- a disperser that sends reflected radiation of different wavelengths in different directions
- a radiation detector that accumulates the dispersed radiation, converting it into a storable form
- a storage device from which the spectrum can be extracted for later examination; this is often in the form of a graph of the radiation collected, versus its wavelength.

In an *emission* spectrum the radiation does not come from some external source but from the distant object itself, and for the comparatively cool temperatures of planetary surfaces/upper atmospheres this will be at infrared wavelengths. Figure 5.1 still applies.

A notional spectrum is shown in Figure 5.2. You can see that the reflected (or emitted) radiation is constant over a small range of wavelengths, then shifts to some other level. In fact, the radiation reflected (or emitted) by a body is a continuous curve, though sometimes with sharp increases and decreases. The constant levels are called bins, the narrower the bin the closer we approach the true curve. In practice there is a limit, for two reasons. First, no disperser can adequately separate arbitrarily close wavelengths – it has a limit to its spectral resolution. This sets a lower limit to the useful bin width. A glass

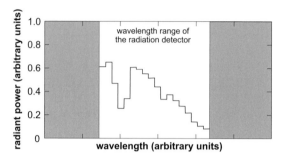

FIGURE 5.2 The notional spectrum of a body, shown as radiant power received versus the wavelength of the radiation. In practice, the sensitivity of the detector is not constant within its wavelength range, but declines towards each extreme.

prism has relatively poor dispersion; much better is a device called a diffraction grating, which consists of many fine lines on a substrate.

Second, the greater the amount of radiation the detector accumulates, the larger and therefore the more reliable its recorded signal. This amount can be increased in four ways:

- by lengthening the exposure time
- by increasing the area of the capturing lens or mirror
- by increasing the sensitivity of the detector
- by allowing a range of wavelengths (from a number of adjacent bins) to fall on a detecting element.

In the case of a PMT there is a single detecting element, whereas in the case of a photographic plate, or its electronic replacement, the CCD, there is an array of closely spaced elements. In both cases, a decision has to be made as to how many adjacent elements, each receiving a different wavelength, will have their outputs added. There is thus a compromise between aggregated bin width, which determines the spectral resolution, and the statistical scatter (noise) in the detector output from the (aggregated) bins.

Figure 5.3 shows the emission spectrum of the Sun (in very narrow bins); it varies slightly on various timescales, but the spectrum shown is typical. You can see that it covers a large range

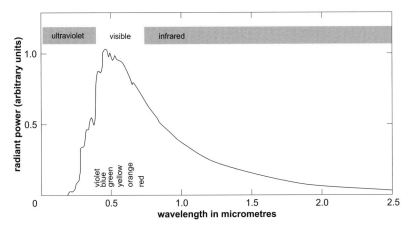

FIGURE 5.3 A typical solar emission spectrum, showing the radiant power versus wavelength in micrometres. The wavelength axis is labelled to show the ranges that correspond to the ultraviolet (UV), the visible and the infrared (IR). At shorter wavelengths there are X-rays then gamma rays, and at longer wavelengths there are microwaves and radio waves. These various ranges are all parts of the electromagnetic spectrum.

of wavelengths, and therefore this spectrum has been obtained by several different spectrometers, each operating over a different wavelength range. This is the true solar spectrum, and not a spectrum distorted by uncorrected variations in detector sensitivity with wavelength, nor by atmospheric absorption. With our eyes we see the Sun only at visible wavelengths, which clearly account for only a modest fraction of the Sun's output of radiation. The three colour sensors in the human eye, when exposed to the Sun's radiation, give the visual sensation of a pale yellow tint.

The reflection spectrum of a body is the fraction of incident radiation at each wavelength that is reflected away. To obtain a Solar System body's true reflectance spectrum allowance has to be made for the Sun's spectrum. For example, if the reflected signal of a wavelength of yellow light from the Sun is twice that of a wavelength of blue light, then an erroneous conclusion is that, at these two wavelengths, the body's reflectance of yellow light is twice that of blue light, i.e. a ratio of 2. But the Sun emits less radiation at blue

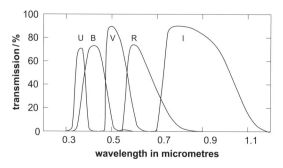

FIGURE 5.4 Wavelength ranges of UBVRI filters – U (UV), B (blue), V (visual – green), R (red), I (near IR).

wavelengths than at yellow wavelengths (Figure 5.3). Therefore the true ratio is less than 2. If the spectrum reflected from a surface is the same as the incident spectrum, then it is called a white surface.

As well as correcting for the source spectrum, adjustments also have to be made for the different sensitivity of the detector to different wavelengths. It cannot be claimed that no radiation is received in the near infrared (just beyond the red) if the detector has zero sensitivity there!

Once a spectrum is obtained, it can be compared with laboratory spectra of known substances. The near infrared (wavelength range 0.8–2.5 micrometres), along with slightly longer wavelengths, is particularly informative.

Standard filters

Short of obtaining a detailed spectrum, coarse spectral information can be obtained by using filters, i.e. substances that transmit only a restricted range of wavelengths. Filters have been used in astronomy for well over a century, but it was only in the early 1960s that an accurate standard set of filters became widely available for visible and near infrared wavelengths. This is the UBVRI set, with transmittances versus wavelength as shown in Figure 5.4.

Filters are normally used one at a time, in which case spectral information is obtained serially, rather than in parallel as with a spectrometer. But regardless of how filters are used, the information we obtain about the composition of a surface or atmosphere of a body is much less than with a high resolution spectrum. To take a homely example, if a surface looks bright through an R filter, but very dim through the others, we would rightly conclude that it has a red tint. But is it a rock rich in iron, a ripe tomato, skin covered in blood or something else?

Let's turn now to the surfaces of Pluto and Charon.

5.2 SURFACES

Pluto

In Section 4.1 you saw that good lightcurves, which were obtained from the mid 1950s onwards, indicated that the reflectivity of Pluto's surface varies across it. High contrast markings are indicated by the range of reflectivity spanned within each lightcurve. After Pluto's size was determined accurately in the late 1980s, an accurate value of Pluto's albedo was determined. It was 55%, indicating that much of the surface is as bright as fresh snow. For comparison, note that the Moon's albedo is only 12%.

In the 1970s Leif Andersson and John Fix concluded from the changing aspect of Pluto as seen from the Earth in the years 1954–1972, that the polar regions are more reflective than elsewhere. This is clearly consistent with polar caps.

But though the albedo, together with the lightcurves, indicate a surface with a large proportion covered by fresh icy materials, perhaps with polar caps, to go any further surface spectra were needed.

It was in 1933 that Vesto Slipher, using colour filters, concluded that Pluto's reflectivity at yellow and red wavelengths is higher than at shorter visible wavelengths, so its surface (on average) has a yellow tint. Alas! As you saw in Section 5.1, tint tells us little about composition. Unfortunately, it was impractical to obtain useful spectral resolution for several decades after Pluto's discovery. Though PMTs had been available since the early 1950s, they could only deliver a single measurement at a time, resulting in impractically long times to build up a detailed spectrum.

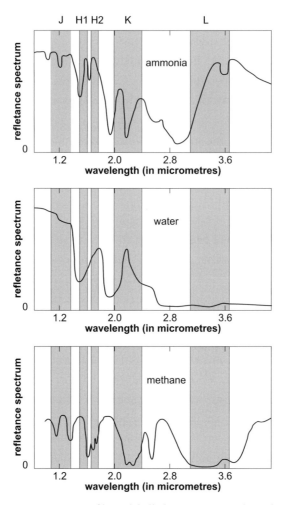

FIGURE 5.5 Five filters, labelled J, H1, H2, K and L, selected to detect ices of ammonia (NH₃), water (H₂O) and methane (CH₄) in the reflectance spectrum of Pluto. Their transmission bands (shaded) are shown superimposed on each ice's reflectance spectrum.

It was in 1976 that Dale Cruikshank, Carl Pilcher and David Morrison obtained the first really useful data on surface composition. Though they used filters, these were narrow band and all in the near infrared, selected to detect some of the expected surface ices, water (H_2O), ammonia (NH_3) and methane (CH_4) (see Figure 5.5). Using the

4-metre reflector Mayall Telescope at Kitt Peak, they deduced the presence of CH_4 frost on Pluto. They did not see rock, H_2O ice or NH_3 ice. Two years later, CH_4 ice was detected by a new generation of infrared spectrometers.

The match obtained by Cruikshank *et al.* between Pluto and laboratory data using the same filters was not perfect, indicating the presence of something else, but the data were insufficient to identify what it was.

The discovery of CH_4 indicated that, even if Pluto had formed closer to the Sun than where it lies today, and then migrated outwards (Section 4.3), it could not have formed a lot closer in; if it had, then the volatility of CH_4 would have resulted in the loss of so much of it that deposition of detectable amounts on the planet's surface would not have occurred. The failure to detect NH_3 and H_2O ices could be put down to their far lower volatilities than that of CH_4; Pluto's high albedo indicates resurfacing by frosts deposited after sublimation, and CH_4 sublimates much more readily than NH_3 and H_2O. Note that sublimation is when a solid evaporates to form a gas, without going through the liquid phase. To melt to form a liquid a certain minimum pressure is required – see Section 5.3.

Table 5.1 indicates the reason for the differences in the volatility of icy materials. Broadly speaking, the lower the melting temperature the more volatile the icy material. As in Table 3.1 the melting temperatures are at the triple point of each substance (Section 5.3). The essential point to take from Table 5.1 is that it shows the correct ordering of volatility, with water as the least volatile and nitrogen as the most volatile.

Without the cycle of sublimation followed by deposition, Pluto's methane would darken in about 100 000 years, solar UV radiation converting CH_4 into much darker hydrocarbon deposits that have much more massive molecules, with a balance set up between darkening and renewal. This balance shifts as Pluto moves around its orbit, which goes part way to explaining Pluto's decreasing albedo since the 1950s up to the mid 1980s. But over these decades the

Table 5.1 *Melting temperatures of common icy materials.*

	Melting temperature	
	(°C)	(K)
Water, H_2O	0	273
Carbon dioxide, CO_2	−56.6	216.4
Ammonia, NH_3	−77.7	195.3
Methane, CH_4	−182.5	90.5
Carbon monoxide, CO	−205	68
Nitrogen, N_2	−210	63

The temperatures are at the triple point.

viewing geometry has changed, from a more pole-on view in the 1950s than subsequently, and, with the polar regions being bright, this has surely contributed to the darkening. Since the mid 1980s no further darkening has been observed. Pluto might also have become redder as it grew darker, a common phenomenon in the outer Solar System, either because of ice thinning and/or because the ice ages.

There was always the possibility that the CH_4 signatures came partly, even wholly, from Charon (which had been discovered in 1978). This was ruled out after observations by several groups in 1987, when central events started (Section 3.1). The signature came wholly from Pluto.

The next major step forward came in 1993. In that year spectra of Pluto were obtained by Tobias Owen (University of Hawaii) and his collaborators, with the then new UK Infrared Telescope (UKIRT) on Mauna Kea, equipped with a very high resolution spectrometer. In a few days, not only was CH_4 ice detected, but also the weaker spectral signatures of nitrogen ice (N_2) and carbon monoxide ice (CO). You might think weaker signatures mean lower abundances, but this is not so. The molecular structure of N_2 gives it an intrinsically weak spectral signature in the infrared. When allowance was made for these intrinsic properties the team deduced that the ices on the *surface* of Pluto were present in the proportions, by numbers of molecules, N_2 ice 98%, CH_4 ice 1.5%, CO ice 0.5%, which would be intimately mixed together. Thus, whereas H_2O ice is the only ice naturally occurring on the surface of the Earth, it seems that N_2 ice dominates the surface of Pluto. However, going from spectra to proportions is fraught with uncertainties, and later work by others indicates that N_2 ice accounts for a considerably smaller proportion of Pluto's surface ices.

H_2O ice, being the most abundant icy material in the Solar System, must be present on Pluto, but at the low temperatures there, at least down to modest depths, it will be so far below its sublimation point that its sublimation rate will be negligible. It will thus lie beneath the more volatile ices.

The proportions of the various ices at the surface will change with Pluto's distance from the Sun. This is because of the decline in surface temperatures with increasing solar distance, which influences the balance between sublimation and deposition differently for different ices.

Maps of Pluto's surface

That the surface of Pluto is patterned had been revealed by Pluto's lightcurves from the 1950s onwards, as you learned earlier. There was also an indication from the 1970s that bright polar caps were present. Further advances were made in 1983, when the lightcurves were used by Robert Marcialis (University of Arizona's Lunar and Planetary Laboratory) to obtain a coarse global map of Pluto. He used a computer model that sought the fewest number of regions that could account for the lightcurves, with the outcome shown in Figure 5.6, the result of lengthy computations. You can see that there are two dark spots, two bright polar caps and an equatorial zone of intermediate brightness. These patterns comprise regions with different reflectivities, presumably relating to the condition of the icy materials at the surface.

FIGURE 5.6 A global map of Pluto obtained from Pluto's lightcurves, by Robert Marcialis in 1983. North is at the top.

Further mapping took place around 1990, using not only the ever increasing number of lightcurves, but the repeated transits of Charon across Pluto's disc from 1985 to 1991, which acted rather like a moving shutter. Only the Charon-facing hemisphere of Pluto could be mapped. Marc Buie (Lowell Observatory) and colleagues, in a sophisticated computer model, used 'tiles' of various reflectivity to cover the surface of Pluto, to determine the configuration that best fitted the lightcurves and the transits. One of the many complications was that a model of Charon's contribution had to be included. They published their map in 1993, more detailed than that in Figure 5.6 and showing only the broadest similarities, namely that the poles are brighter than the equatorial regions and that the latter are of uneven brightness. At about the same time, Eliot Young and Richard Binzel, both at MIT, using the Charon transits only, and a different computer model, obtained a map of the hemisphere of Pluto that faces Charon. Though there are differences in the two maps of this hemisphere, there are commonalities, notably

- reflectivities ranging from 15% (rather dark) to over 70% (like fresh snow) – a large range
- one, perhaps two, polar caps
- dark bands and bright spots.

In 2001 Eliot Young, Richard Binzel and Keenan Crane, used data from the same transits, and in particular the central events that lasted from 1987 to 1989 (when the whole disc of Charon passed between us and Pluto's disc). From B and V filter data (Figure 5.4), they produced the two-colour map shown in Figure 5.7. The bright

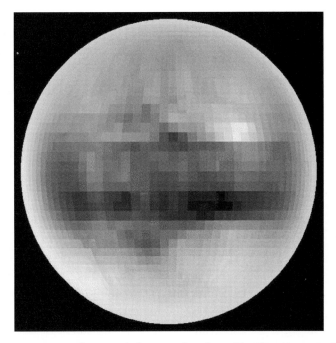

FIGURE 5.7 The map of Pluto's surface derived by Eliot Young, Richard Binzel and Keenan Crane in 2001, using Charon's transits of Pluto. North is at the top. Only the Charon-facing hemisphere could be mapped. (See plate section for colour version.)

areas are primarily N_2 frost with traces of CO and CH_4 frosts. The composition of the dark areas in unknown, but is presumed to be the result of the action of solar UV radiation on CH_4 frost. The variations within the dark regions could either be due to bright frost interspersed with dark material in a non-uniform way, or variation with location of the degree of UV darkening of the ices.

Figure 5.8 shows two surface maps of Pluto. The upper one was obtained from images taken in 1994 by ESO's Faint Object Camera on the HST. There is certainly one polar cap, plus bright and dark patches in the equatorial region. The lower map was obtained from images taken in 2002–2003 by the HST's Advanced Camera for Surveys. It

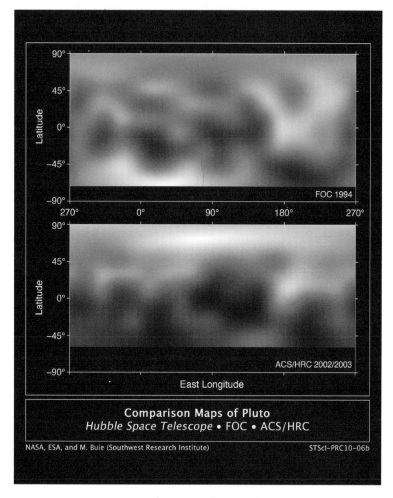

FIGURE 5.8 Two surface maps of Pluto. The upper one was obtained from images taken in 1994 by ESO's Faint Object Camera on the HST. The lower map was obtained from images taken in 2002–2003 by the HST's Advanced Camera for Surveys. Released by Marc Buie in 2010.

also shows a polar cap, now at the other pole, plus (different) bright and dark patches in the equatorial region.

It was noted earlier that the proportions of the different ices on Pluto's surface must vary with solar distance. This is surely the case for surface patterns too, again because of the decline in surface temperatures with increasing solar distance, which influences the balance between sublimation and deposition differently for different ices. The maps in Figure 5.8 are certainly in accord with this expectation.

Furthermore, during 2002–2003, the images acquired showed a gradual increase in the red tint of Pluto's surface, and what is thought to be N_2 ice growing and shrinking, getting brighter in the north and darker in the south.

Note that the Advanced Camera for Surveys is now defunct. The lower map in Figure 5.8 is the best resolution we'll have until the arrival of *New Horizons* in 2015.

Whether the reflectance features relate closely to topography is unknown – we know nothing about the topography of Pluto.

What of the sources of the ices on the surface of Pluto? These could be primordial, dating back to the formation of Pluto, or they could be refreshed by icy volcanism (cryovolcanism) bringing up icy materials from the interior – more on this in Section 5.4.

Charon

After Charon's size was obtained accurately in 2005, when it occulted a star, its albedo could be determined. This was found to be 35%, considerably less than Pluto's 55%, indicating that whatever icy materials covered Charon they were not as fresh as those on Pluto. This was the first indication that Charon had little atmosphere through which a sublimation-precipitation cycle could be established. This was hardly surprising for a body that was known to have a mass considerably less than even Pluto's small mass. Charon could not have held on to a significant atmosphere.

During the epoch of the central events in the occultation of Charon by Pluto and the transit of Pluto by Charon, spectra were obtained for Pluto plus Charon and, with Charon wholly behind Pluto, Pluto only. The difference gives Charon only. Figure 5.9 shows the three spectra obtained in 1987 by Uwe Fink and Michael DiSanti (University of Arizona). As well as showing no CH_4 feature on Charon (a result referred to earlier), it showed that Charon's spectrum was

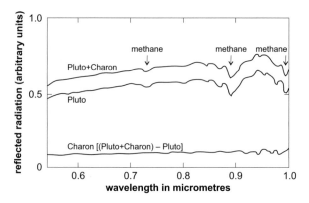

FIGURE 5.9 Reflection spectra obtained by Uwe Fink and Michael DiSanti in March 1987, showing the methane (CH_4) features that come from Pluto alone, and the featureless, flat spectrum of Charon.

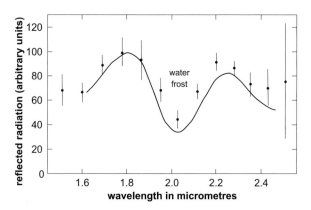

FIGURE 5.10 The infrared spectrum of Charon, obtained in April 1987 by Marc Buie *et al.*, showing the water (H_2O) frost absorption feature. The vertical lines centred on each data point denote uncertainties in the measurement.

featureless, and fairly flat. This means that Charon's visual appearance is grey, not red tinted like Pluto.

Figure 5.9 extends only a little way into the near-IR. Further into the near-IR Robert Marcialis with George and Marcia Rieke, detected absorption features of water (as frost). During the 23 April 1987 central event this was confirmed by Marc Buie and colleagues on Mauna Kea; their spectrum is shown in Figure 5.10. Thus, even though CH_4 ice

FIGURE 5.11 A map of Charon, based on the Pluto-Charon transits/ occultations that occurred from 1985 to 1991. (Marc Buie)

was absent at Charon's surface, the much less volatile H_2O ice was there. In 2000, spectra of Charon were obtained by Michael Brown and Wendy Calvin (California Institute of Technology), and also by Marc Buie and William Grundy (Lowell Observatory), that showed that most of the H_2O ice is crystalline, i.e. the water molecules are arranged in a regular repeating pattern. The rest of the H_2O ice is amorphous, i.e. the H_2O molecules are jumbled. In 2007 it was estimated that over 90% of the H_2O ice is crystalline. This has significance for the interior of Charon, so I'll return to the topic of H_2O ice on Charon in Section 5.4.

NH_3 ice and ammonia hydrate ices have also been detected spectroscopically, initially in 2000 by Brown and Calvin, and Buie and Grundy. These ices are also less volatile than CH_4 ice. Other substances, rocky and icy, could be present, but there is no evidence yet.

Charon's lightcurve range, lightest to darkest, was measured from HST images to be 8% in 1993, which is several times less than Pluto's 1992 HST value of 38%. This shows that Charon's surface is the more uniform. Nevertheless, Marc Buie used the Pluto-Charon transits/occultations that occurred from 1985 to 1991 (Section 3.1) to obtain the reflectivity map of Charon in Figure 5.11. However, it

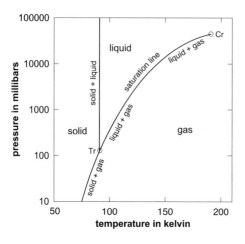

FIGURE 5.12 The phase diagram of pure methane, CH₄, which shows the pressure and temperature ranges over which CH₄, in an enclosed environment, is in equilibrium as a solid, a liquid or a gas. Note that along the pressure axis the pressure increases ten times from one tick mark to the next. This is a logarithmic scale, used to accommodate a large range of values.

is difficult to identify Charon's contribution to the data, and consequently much of the detail in Figure 5.11 is spurious; only the edges might have some validity.

I now turn to the atmospheres of Pluto and Charon.

5.3 ATMOSPHERES

Your understanding of planetary atmospheres will be significantly enhanced if you read the following material on what are called phase diagrams. I've reduced the material to the minimum necessary for a basic understanding of what determines whether a substance will be present as a solid, a liquid or a gas.

Phase diagrams

I'll use methane's phase diagram as an example, a very relevant one at that. Figure 5.12 shows the phase diagram of pure CH₄. A phase diagram shows the pressure and temperature ranges over which a substance in an enclosed environment is in equilibrium as a solid, a liquid or a gas. 'Phase' is the generic term for solid, liquid and gas. Pressures are in millibars, thousandths of a bar, where 1000 millibars (1 bar) is very close to the mean atmospheric pressure at sea level on the Earth. Note that temperatures are in kelvin, K (same as the Celsius scale except that zero, absolute zero, is at $-273\ °C$).

A phase diagram displays the boundaries between the phases. If the pressure and temperature of a substance are *anywhere* within the area to the left of the solid+gas or solid+liquid boundaries, it will be solid. Unsurprisingly, the solid phase is confined to the lower temperatures. On each phase boundary just two phases coexist – the phases that meet at the boundary. The liquid phase requires a minimum pressure and temperature, marked by Tr in Figure 5.12. At pressures below Tr a substance cannot be in the liquid phase at *any* temperature. For pure CH_4 the temperature at Tr is 90.4 K and the pressure 131 millibars. Tr marks what is called the triple point because it is where all three phases co-exist in equilibrium.

At pressures above Cr, which marks what is called the critical point, there is no distinction between a gas and a liquid. For example, if, in this region, at constant pressure, the temperature is increased from a modest value where the density of the fluid is liquid-like, to a high value where the density is gas-like, there would have been no sharp change in density. By contrast the phase boundaries mark where there *is* a sharp change: a large decrease from liquid to gas and from solid to gas, and a much smaller decrease from solid to liquid. (Water is a *very* rare exception, where there is a slight *increase* in density from solid to liquid.)

The various phase changes have specific names, as follows: solid to gas – sublimation; gas to solid – deposition; solid to liquid – melting; liquid to solid – freezing; liquid to gas – evaporation; gas to liquid – condensation. Note that temperatures on the solid+liquid line are *far* less sensitive to pressure than on the other two lines.

The pressure on the solid+gas or solid+liquid boundaries is called the vapour pressure of a substance. It decreases a great deal for a small fall in temperature. In the case of CH_4 you can see from Figure 5.12 that a ten-fold decrease in vapour pressure from 100 millibars to 10 millibars corresponds to a temperature decrease of only about

16 K. You can also see that the solid+gas line steepens as temperature declines. Below 75 K the vapour pressure sensitivity to temperature soon becomes huge, decreasing very rapidly as temperature declines. At 40 K the vapour pressure is around 25 *millionths* of a millibar!

Pluto: the early years

It was in 1943–1944 that the first major attempt to detect an atmosphere on Pluto was made. This was by Gerard Kuiper, using a photographic spectrometer on the 2.08-metre (82-inch) reflector at McDonald Observatory. He searched for CH_4, reasoning that it is relatively abundant in the Solar System, would be gaseous even in the cold outer Solar System, and gives a strong spectral signature even in small amounts. He failed to detect CH_4 on Pluto but, given the modest sensitivity of his photographic spectrometer, this did not rule out a tenuous CH_4 atmosphere.

The 1976 identification by Cruikshank and his colleagues of CH_4 in their spectra of Pluto established that CH_4 is present, but it was not known how much of this was due to surface ice, and how much to atmospheric gas; much higher resolution infrared spectrometers were needed to find out, and these were not available until 1994.

The vapour pressure could be obtained from the phase diagram of CH_4, which will be the total atmospheric pressure if no other gases are present. But you have seen that this pressure is very sensitive to the methane ice temperature. The surface temperature up to the early 1980s had to be calculated (see Box 5.1), but the size and radiative properties of Pluto were not well known. This resulted in estimates in the range 40–65 K, corresponding to a very wide range of vapour pressures. At 65 K the vapour pressure would be about 0.9 millibar, and of order 25 millionths of a millibar (25 nanobars) at 40 K – atmospheric pressure at the surface of the Earth is around 1000 millibars. Clearly, CH_4 cannot give Pluto much of an atmosphere! However, the bright areas that account for Pluto's high albedo indicate an atmosphere of some sort as noted in Section 5.2, to maintain a refreshing sublimation-deposition cycle. That Pluto has an

BOX 5.1 CALCULATING SURFACE TEMPERATURES
(FOR THOSE COMFORTABLE WITH ALGEBRA)

The surface temperature of a planetary body depends on the balance between the solar radiation it absorbs and the radiation it emits to space. The solar radiation *absorbed*, W_{abs}, is given by

$$W_{abs} = F_{solar} A_p (1 - a_B),$$ (1)

where F_{solar} is the radiant power per unit area in the solar radiation at the distance of the planetary body from the Sun. A_p is the projected area of the body's surface presented towards the Sun. The factor a_B is the Bond (or planetary) albedo, i.e. the fraction of the intercepted solar radiation that is reflected back to space by the surface and any atmosphere (William Cranch Bond, US astronomer, 1789–1859). Do not confuse this with the albedo introduced earlier, which is the radiation reflected towards the observer compared with that received from a standard surface (a Lambertian surface).

L_{out} is the radiant power emitted by the warmed body to space (at infrared wavelengths). If there is negligible heat generated within the body, which is an excellent approximation for a small body like Pluto, then, in equilibrium,

$$L_{out} = W_{abs}$$

and using Equation (1)

$$L_{out} = F_{solar} A_p (1 - a_B).$$ (2)

It is usual to express L_{out} as the effective temperature of the planetary body. This is defined in terms of the radiation from what is called an ideal thermal source – in essence it is a surface that absorbs all the radiation that falls on it. The power radiated per unit area of such a source depends only on its absolute temperature, T, and is given by σT^4 where σ is a universal constant called Stefan's constant. For an ideal thermal source of surface area A the power radiated is therefore given by

$$L_{out} = A(\sigma T^4).$$

If A is the *total* area of a planetary body (not the projected area) then this equation *defines* T to be the effective temperature T_{eff} of the planetary body.

Thus, *by definition*, $L_{out} = A(\sigma T_{eff}^4)$, which can be rearranged as

$$T_{eff} = [L_{out}/(A\sigma)]^{1/4}.$$

Using Equation (2) to substitute for L_{out} this becomes

$$T_{eff} = [F_{solar} A_p (1 - a_B)/(A\sigma)]^{1/4}$$

For a spherical body of radius R, $A_p = \pi R^2$ and $A = 4\pi R^2$, therefore

$$T_{eff} = \left(\frac{F_{solar}(1 - a_B)}{4\sigma} \right)^{1/4}. \tag{3}$$

This is useful in that T_{eff} depends only on the radiant power per unit area in the solar radiation at the orbit of the planetary body, and on its Bond albedo – the radius of the body has been eliminated.

However, a planetary body does not radiate in the manner of an ideal thermal source, which raises the question of what T_{eff} means in terms of actual temperatures. For a body without an atmosphere the global mean surface temperature T_s will be approximately equal to T_{eff} if the surface is a very good absorber of radiation at the wavelengths of its emission (which will be at IR wavelengths). The less good it is the greater will be T_s for a given T_{eff}. For a body with a substantial atmosphere, the radiation emitted to space comes from a range of altitudes in the atmosphere, as well as from the surface. In this case, T_{eff} is the temperature at some altitude above the surface, but this is not from where all the emission comes.

Clearly to obtain T_s in this way a lot has to be known or guessed about the planet.

atmosphere was put beyond doubt by a rare stellar occultation in June 1988. The star did not disappear and re-appear instantaneously, but gradually.

Pluto: the 1980s and 1990s

The astronomical community was well prepared for the 1988 occultation, with observers spread out over the zone from which it would be visible, a band of latitude that included most of Australia, plus a few observers just outside this zone. In addition to ground-based telescopes, the 0.91-metre reflector on the Kuiper Airborne Observatory (KAO) was used.

Observers inside the zone saw the star disappear rapidly, but not instantaneously – only if Pluto had no atmosphere would the latter have been the case. It is the slight bending of the light from the star as it passes through the atmosphere – refraction – that causes the more gradual disappearance. This refraction can be used to obtain the density of the atmosphere versus altitude. The surface density, combined with the surface temperature (then only known to be somewhere in the range 40–65 K), gives the surface pressure – values in the range 0.003–0.01 millibars were obtained. Clearly CH_4 comes nowhere near being able to account for such a comparatively high pressure. What other gases were present?

Following the 1988 occultation, Roger Yelle (Boston University) and Jonathan Lunine (University of Arizona) modelled the thermal profile and fitted the data with an atmosphere consisting of a small amount of methane but largely of a gas with a molecular weight of 28 (a total of 28 protons and neutrons). The common isotope of molecular nitrogen, N_2, is just such a gas with 14 protons and neutrons in the nucleus of each of its two nitrogen atoms. In 1993 Tobias Owen (University of Hawaii) confirmed that the gas was indeed N_2.

It is to be expected that the composition of Pluto's atmosphere is determined by its surface composition and the volatility of each surface component. You have seen that the surface of Pluto is coated in N_2, CH_4 and CO ices, and that N_2 ice is much the most volatile. The

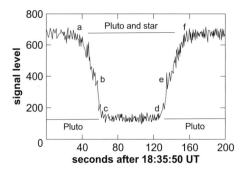

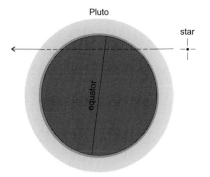

FIGURE 5.13 The lightcurve obtained by the Kuiper Airborne Observatory when Pluto occulted a star in June 1988. The inset shows the path of the star relative to Pluto; note that the star is essentially fixed in space, and that it is Pluto that moves with respect to the stars, from left to right. Note that the atmosphere does not have the hard upper edge shown here, but thins very gradually.

atmospheric composition is clearly in accord with this. N_2 must be the dominant atmospheric constituent, with CH_4 and CO present in far smaller quantities. That CH_4 was detected in the atmosphere first is because of its strong spectral signature in the infrared. Atmospheric composition also depends on the distribution of ices. For example, patches rich in CO and in CH_4 are known to exist, which would boost the atmospheric abundance of these two constituents over the case when they are intimately mixed with other ices.

Figure 5.13 shows the 1988 occultation result from the KAO. As well as the gradual disappearance and emergence of the star, a steepening of the lightcurve is apparent. Two possible explanations were advanced. James Elliot (MIT) and Leslie Young (Southwest Research Institute) suggested a substantial layer of haze, whereas Von Eshelman (Stanford University), William Hubbard (University of Arizona) and

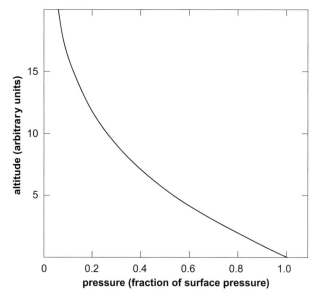

FIGURE 5.14 A schematic representation of the decline in atmospheric pressure with altitude above the surface of a planet.

colleagues suggested a steep increase in temperature with decreasing height at low altitudes, which would influence the refraction by effecting atmospheric density. This increase could result from the greenhouse effect produced by CH_4, i.e. the trapping in the lower atmosphere of the thermal radiation emitted by Pluto. Of the two explanations, the haze layer is preferred. A *substantial* haze would result in an overestimate of Pluto's radius from Charon's transits of Pluto, though by no more than about 50 km. This would also apply to the stellar occultation. An impenetrable haze means a deeper atmosphere, perhaps about 0.05 millibars at the surface.

Whatever the surface pressure, be it as low as 0.003 millibars or as 'high' as 0.05 millibars, one thing is entirely clear from Figure 5.13: Pluto's thin atmosphere extends a long way from Pluto's surface. Atmospheric pressure declines as altitude increases, tailing off into space, as shown schematically in Figure 5.14, so some reasonable definition of atmospheric thickness is needed. The one adopted is the scale height. This is the increase in altitude necessary to reduce the pressure by a factor 2.718... This curious number, which, like π, goes on for ever, is denoted by e, and emerges in mathematics in relation to exponential growth and decay.

For the Earth the scale height in the lower atmosphere is about 40 km, whereas for Pluto it is about 60 km. This greater value, bearing in mind Pluto's diameter of only about one fifth that of the Earth, means that, relative to our atmosphere, Pluto's atmosphere is far more distended. This, coupled with Pluto's lower gravity, which at the surface is about 8% of that at the Earth's surface, means that there is a significant loss of atmosphere to space, particularly around perihelion when surface and atmosphere are at their least cold. This loss is replaced by sublimation from the surface. Up to about 30 km depth of ice could have been lost to space over Pluto's 4600-million-year lifetime.

Attempts in the 1980s and 1990s to obtain Pluto's surface temperature have an intricate history, well documented by Stern and Mitton (see Further *reading and other resources*). I'll go forward to 1997 when the powerful Infrared Space Observatory (ISO), in Earth orbit, measured the IR flux from Pluto as the planet rotates, thus displaying to us hemispheres with various proportions of light-to-dark regions. The derived temperatures showed that the bright areas are at 40 K, and the dark areas are at 60 K. Note that these are daytime temperatures. This is because Pluto is so far from the Earth that we see it never more than a couple of degrees from full; it's as if we only ever saw a (near) full Moon in our skies. The dark side of Pluto, of which we see at most a thin sliver, would be far colder, perhaps of order 10 K. ISO could not resolve Pluto and Charon, so the IR flux contributed by Charon had to be removed by modelling.

That the dark areas are warmer is a result of their absorbing more solar radiation, whereas light and dark areas do not differ greatly in their loss of heat to space. Were clouds present in Pluto's atmosphere this would make the determination of surface temperatures extremely difficult. Fortunately, this possibility was ruled out by the smallness of the variations in the shape of Pluto's lightcurve over many rotations; clouds would have to have invariant properties and locations for such a result, and that's not the case for clouds.

Pluto's eccentric orbit has perihelion at 29.7 AU and aphelion at 49.6 AU. Far from perihelion the surface of Pluto will be much colder than even its frigid surface is today, so much so that even much of the highly volatile N_2 will precipitate on to the surface. Consequently, the atmospheric mass will decrease. Will there be any left when the spacecraft *New Horizons* (Chapter 8) reaches Pluto 1n 2015? Read on.

Pluto: the twenty-first century
The decline in the mass of Pluto's atmosphere as it recedes well beyond its 1989 perihelion in its 249 year orbit has been calculated. The result depends on uncertain things such as the distribution of reflectivities on Pluto, and the distributions of each of the three ices over its surface. Predictions vary from a sudden collapse over a few years to a decades-long decline, with timings of the onset in each case varying from the mid 1990s to the 2030s. In 2002 two stellar occultations occurred, the first since 1988. Surprisingly, the atmospheric pressure had doubled! This is good news for *New Horizons*. Also, the haze layer/temperature gradient change high in the atmosphere had moderated. The increase in pressure *after* perihelion is presumably due to the non-instantaneous response of the surface ices to changes in the radiation received from the Sun.

On 12 June 2006 there was another occultation, which occurred shortly after Pluto began its ten-year passage in front of the region of the galactic centre (in Sagittarius), which is rich in stars. James Elliott and colleagues observed the occultation from five sites in south east Australia. High in the atmosphere the temperature was about 100 K, increasing slightly as altitude decreased, but lower in the atmosphere the temperature suddenly became more sensitive to altitude, decreasing by about 2.2 K per km. Much the same thermal structure was seen in 2002, but in 1988 a much larger thermal gradient was inferred in the lower atmosphere, though more haze in 1988 is the preferred possibility, as you saw earlier. The estimates of Pluto's diameter from the 2006 occultation have been discussed in Section 3.1.

The next stellar occultation was on 18 March 2007. Astronomers obtained data using six observatories in the western USA, including the 6.5-metre Multiple Mirror Telescope on Mount Hopkins in Arizona. The atmospheric structure was much the same as in 2006. It was also confirmed that the increase in atmospheric pressure measured between 1988 and 2002 had ceased. However, it is predicted that a substantial atmosphere will certainly be present when *New Horizons* arrives in 2015. At present (2009) the surface pressure on Pluto is somewhere in the range 0.0065–0.025 millibars.

Aside from occultations, in 2005 the submillimetre radiotele-scope array (SMA) on Hawaii was used by M A Gurwell and B J Butler to establish the average daytime surface temperatures on Pluto and Charon – the SMA could resolve them. For Pluto the result is 40 ± 4 K, and for Charon 56 ± 14 K, The value for Charon is more or less in accord with an atmosphere-less surface of known albedo. By contrast Pluto is cooler than had been predicted. Why? A plausible explanation is that some of the solar radiation is being taken up in vaporizing N_2 ice at the surface, leaving less radiation to warm the surface.

The story continues, and in 2009 E Lellouch (Paris Observatory) reported the results of the group he led to observe Pluto with one of the four 8-metre telescopes that constitute the European South-ern Observatory's Very Large Telescope in Chile. A cooled infrared spectrograph was used (CRIRES). Upper atmospheric temperatures were much the same as found earlier, around 100 K. A new result was the measurement of the amount of CH_4 gas in the atmosphere, 0.5 ± 0.1 %, much the same proportion as at the surface. The authors conclude that it is this CH_4, derived from CH_4 frost at the surface, which warms the atmosphere through its efficient absorption of IR radiation from the surface, i.e. through its greenhouse effect, raising the surface temperature above what it would otherwise be.

The Spitzer infrared space telescope, more sensitive than ISO, has observed Pluto and has detected temperature variations across the surface. Detailed results are awaited.

Charon

In Section 5.2 you saw that spectra of Charon have revealed H_2O ice to be present on the surface, along with ices of NH_3 and ammonia hydrates. These are far less volatile than the ices that cover Pluto (N_2, CH_4 and CO). Also, Charon's lower gravity makes it less able to retain an atmosphere. Therefore, not much of an atmosphere can be expected.

During the rare stellar occultation of 11 July 2005 (Section 4.2) no atmosphere was detected, the star just went out like a light! If Charon's atmosphere is pure N_2 then the upper limit on the surface pressure is 110 nanobars, and if it is a pure CH_4 atmosphere then it is 15 nanobars.

The best value of Charon's surface temperature has been obtained with the Multiband Imaging Photometer (MIPS) on the Spitzer infrared space telescope, in September 2004. Temperatures in the range 54–59 K were obtained. This too is a dayside temperature, because Charon, like Pluto, is always seen from Earth at or near to full. Also like Pluto, the nightside temperature will be of order 10 K.

Note that as Pluto moves around its orbit, the dayside hemisphere that it presents to us slowly changes. The same applies to Charon. We thus see more or less of the bright regions. Consequently, it is possible that the surface temperatures will exhibit modest changes

Table 5.2 presents the main properties of the atmospheres of Pluto and Charon today.

5.4 INTERIORS

Pluto

In Section 3.3 you saw that Pluto's mean density, 2030 kg/m^3, implies a mixture of icy materials and partially hydrated rocky materials, plus perhaps a small amount of Fe-Ni, with about a third of the mass being of icy materials, particularly H_2O, the most abundant icy material in the Solar System. How are the various materials distributed in the

Table 5.2 *Surface pressures, surface temperatures and compositions of the atmospheres of Pluto and Charon today.*

	Surface pressure[1]	Surface temperature[2]	Composition
Pluto	In the range 0.0065–0.025 millibars	Regional, about 40–60 K, average around 43 K	Mainly N_2 traces of CH_4 and CO.
Charon	Upper limits: 110 nanobars if pure N_2, 15 nanobars if pure CH_4	54–59 K	Undetected

[1] A millibar is a thousandth of a bar, and a nanobar a millionth of a millibar. Mean atmospheric pressure at sea level on Earth is 1013 millibars.

[2] The unit is the kelvin, the same as °C except that its zero is absolute zero, −273°C.

interior? Are they all mixed together like the ingredients of a cake, or are there layers, shells, surrounding a central core?

A layered body is said to be differentiated, and in a planet this can come about in two ways. Either the planet could have built up first from one material, then from another, or it could have been formed homogeneously, and then the intrinsically denser materials settled towards the centre. It is also possible to have a combination of the two, with, for example, a homogeneous core overlaid by an initially heterogeneous mantle that subsequently differentiated.

If a planet is not born differentiated then what determines whether it subsequently undergoes differentiation? The answer is heating. If the interior is warm enough then the intrinsically denser materials will settle downwards, like stones sinking in treacle. Materials don't have to melt for this to happen as long as they are sufficiently warm to undergo plastic flow. Heating results from the formation of a planet, the gravitational energy being converted into kinetic energy as materials come together, with heating when fragments collide. Some of this heat is radiated to space but, particularly if the coming together – the accretion – is rapid, some is trapped and raises the internal temperatures. Subsequently, further heating can arise in various ways, such as from the fast atomic particles and gamma rays emitted by unstable (radioactive) isotopes in rocky materials. These particles and gamma rays are absorbed by the surrounding material, thereby heating it.

Among the Solar System planets, you saw in Section 1.5 that the giant planets, Jupiter, Saturn, Uranus and Neptune, are thought to have formed in two main stages: from a kernel of rocky plus icy materials which then captured gases from the nebular disc, notably hydrogen and helium, to form a dense gaseous envelope. Subsequently some layering developed in this envelope. By contrast, the terrestrial planets, Mercury, Venus, the Earth and Mars, formed much more homogeneously (though an acquisition of volatile materials probably occurred towards the end of formation). There is incontrovertible evidence that the terrestrial planets are now differentiated. Initially the evidence came from thermal modelling, but subsequently from gravity and other measurements at the surface, and then from orbit, gravity measurements by spacecraft being of particular importance.

So, is Pluto differentiated? Models based on its global composition suggest that it might be, and that it might not be! Because of its small size, Pluto is a marginal case, and the issue will only be resolved when the spacecraft *New Horizons* gets to Pluto in 2015. However, it must be noted that a differentiated Pluto is required for the impact origin of Charon outlined in Section 4.3, which accounts for the lower density of Charon by its collisional formation from the outer regions of Pluto.

Apart from the global mean density, there are few observational constraints on the internal structure of Pluto. Figure 5.15 shows one model, which assumes that Pluto is differentiated, which is the preferred possibility among astronomers. This is based on the global

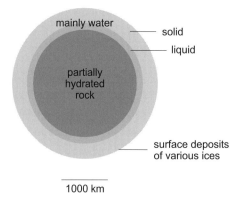

mainly water

solid

liquid

partially
hydrated
rock

surface deposits
of various ices

1000 km

FIGURE 5.15 A model of Pluto's interior, assuming the planet to be differentiated. It is a plausible model, given the weak observational constraints. The partially hydrated rocky core is overlaid by an icy mantle of water (H_2O), liquid and solid, with small quantities of other ices.

mean density, the likely constituents, and the way materials would behave in the interior. The model shown is by William McKinnon of Washington University.

In the model, H_2O ice accounts for about a third of Pluto's mass, constituting an outer mantle, and partially hydrated rock accounts for the rest, constituting a core. There are also traces of denser Fe-Ni in the core, and traces of other ices in the mantle, notably NH_3, which makes it possible that deep in the mantle the mixture is liquid. *If* the ratio of H_2O to NH_3 is 2:1 then the NH_3 depresses the freezing point of the mixture to 176 K, enabling radioactive heating to liquefy the mixture in the warmer, deeper mantle. Note that the high 2:1 ratio need only be present at such depths, and indeed would be extremely unlikely throughout the mantle. Note also that any liquid will be less dense than the overlying solid mantle, so *if* the solid mantle could crack right through, geysers would refresh the surface. There is no evidence that this has occurred, though it has certainly occurred on Triton (Section 4.3).

Table 5.3 gives the densities of some planet-building solids. Note that the densities of these solids are at low pressures, though the values are not very pressure sensitive, and for the materials deep in Pluto's interior will be raised only a little above the tabulated values. Liquids are only slightly more compressible.

Table 5.3 *Densities of some planet-building solids.*

	Density (kg/m^3)
Icy materials	
Water, H_2O	996
Carbon dioxide, CO_2	1980
Ammonia, NH_3	820
Methane, CH_4	494
Carbon monoxide, CO	1240
Nitrogen, N_2	1490
Rocky materials	
Common Earth rocks	2600–3300
Hydrated rocks	~1000–2300
Carbon-rich (e.g. tars)	~1000–2300
Iron with 6% nickel (Fe-Ni)	7925

That's about as far as we can go, until *New Horizons* flies by on 14 July 2015.

Charon

In Section 4.3 you saw that Charon's mean density, 1630 kg/m^3, like Pluto, implies a mixture of icy and rocky materials, with Charon's lower mean density indicating a somewhat higher proportion of icy materials, again with H_2O predominant.

Whether Charon is differentiated is even more uncertain than in the case of Pluto. Figure 5.16 shows a typical differentiated model, the icy mantle being predominantly H_2O ice but with small quantities of other ices, notably NH_3. Jason Cook (Arizona State University) has proposed that there is a 15 km deep water-ammonia ocean at the base of the solid icy mantle.

You saw in Section 5.2 that H_2O ice is widespread on Charon's surface, and that over 90% of it is crystalline. Crystalline H_2O ice (also amorphous H_2O ice) and various ammonia hydrates were detected in

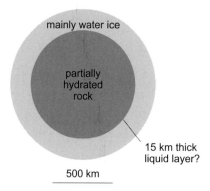

mainly water ice

partially
hydrated
rock

15 km thick
liquid layer?

500 km

FIGURE 5.16 A model of Charon's interior, assuming the satellite to be differentiated. It is a plausible model, given the weak observational constraints. The partially hydrated rocky core is overlaid by a solid icy mantle of water ice with small quantities of other ices.

Charon's near-IR spectrum in 2007 by Jason Cook and his colleagues, using the 8-metre Gemini telescope on Mauna Kea (with adaptive optics to get rid of contamination by Pluto's IR radiation). Crystalline materials are rapidly disordered into amorphous form by cosmic rays, so must be being replenished. How?

Crystalline H_2O ice needs temperatures greater than 78 K to form, yet Charon's surface temperatures are below 60 K. So how could replenishment be accomplished? Three possible means have been suggested by Cook and his colleagues:

- The temperature of amorphous H_2O ice is held above 78 K in the interior, allowing it to anneal and become crystalline. This could then appear on the surface through internal convection.
- It could also be created or brought up by the impacts of small bodies.
- Icy volcanoes – cryovolcanism – could be emitting crystalline H_2O ice.

The first two solutions are not favoured. First, little is known about the possibility of internal convection in Charon. Second, the fresh crystalline ice revealed by impacts is converted to the amorphous form by cosmic rays in a time considerably less than the average time between impacts. The same applies to impact vapour deposited in crystalline form, and to crystalline water ice produced from amorphous ice by the heat from impacts.

The possibility of cryovolcanism remains. *If* Charon contained enough NH_3 when it formed (at least 15% of the icy mantle, which is rather unlikely) then radioactive heating would have sustained a liquid layer at the base of the mantle between the (possibly hydrated) rocky core and the solid icy mantle. In one specific model (Figure 5.16) the ocean is 15 km deep, and has somehow acquired NH_3 to the extent that it consists of H_2O and NH_3 in the ratio 2:1. As you saw in the case of Pluto, the NH_3 depresses the freezing point of such a mixture to 176 K, enabling radioactive heating to sustain it. As Charon cools (over geological times) the liquid freezes and expands; this unusual behaviour of expansion rather than contraction on freezing is due to the H_2O. The overlying ice cracks on an hour time-scale and water gushes forth from the deep interior, and freezes rapidly to create small ice crystals. From Charon's estimated cooling rate, the ocean freezing would currently be resurfacing Charon at a rate of 2 mm per 1000 years, much faster than the rate at which cosmic rays convert H_2O ice to its amorphous form.

The clincher would be images of cryovolcanic eruptions, though unlike cryovolcanism on some of the large satellites of Neptune, notably Triton (Section 4.3), images of such eruptions on Charon are far beyond the reach of our current telescopes.

Unfortunately, the satellites of Uranus comparable in size to Charon are very cold, have surfaces consisting mostly of crystalline H_2O ice, but the surfaces are ancient! The jury on the preservation of crystalline H_2O ice on satellite surfaces remains out – there is no satisfactory explanation as yet.

Doubtless, much more will be learned about Charon during the flyby of *New Horizons* in 2015.

6   The Edgeworth-Kuiper belt

With the discovery of Pluto, was the Solar System complete? No, far from it!

6.1 WHY SEARCH FOR MORE TRANS-NEPTUNIAN OBJECTS?

In 1930, soon after Pluto's discovery, the American astronomer Frederick C Leonard (1896–1960) wondered whether Pluto was the first of many trans-Neptunian objects awaiting discovery, but he does not seem to have acted on his prescient speculation. In July 1943, in the *Journal of the British Astronomical Association*, the Anglo-Irish polymath Kenneth Essex Edgeworth (1880–1972) stated that beyond Neptune the solar nebula was too thinly dispersed to have made planets, but instead many smaller bodies were present, some of which become comets. He expressed similar views in 1949 in an edition of the *Monthly Notices of the Royal Astronomical Society* (*MNRAS*). This was a year before the Dutch astronomer Jan Hendrick Oort (1900–1992) proposed the existence of a distant spherical shell of small icy bodies that enveloped the Solar System and supplied long period comets to its inner regions – the Oort cloud (Section 1.2). Such a cloud is not the belt of trans-Neptunian objects that Edgeworth had proposed; his belt was not so far away and was largely confined to the planes of the planetary orbits.

Whether there was such a belt of trans-Neptunian objects was also considered by the Dutch-American astronomer Gerard Kuiper (1905–1973) in a 1951 edition of the journal *Astrophysics*, but he concluded that such a belt no longer existed! This was based on the then belief that Pluto was about the mass of the Earth and would thus have scattered the bodies away.

FIGURE 6.1 Left: Kenneth Essex Edgeworth, and right: Gerard Kuiper, who each proposed a belt of small objects beyond Neptune in the mid twentieth century. (Edgeworth: Royal Signals Museum, by permission. Kuiper: Special Collections Research Center, University of Chicago, by permission)

Though Edgeworth (in 1943 and 1949), and Kuiper (in 1951) had noted the sharp edge to the Solar System at 30 AU, with only tiny Pluto beyond Neptune, the existence and fate of a distant debris field presumed to have been left over from planetary formation were issues that did not command much interest at the time. In any case, discovering such objects was well beyond contemporary instrument capabilities.

The existence of such a debris field occupying a belt beyond Neptune is consistent with the widely accepted solar nebular theories (Section 1.5). Recall that the nebula was a thick disc of gas and dust which, in its inner regions, gave birth to the planets. It extended well beyond Neptune and, therefore, even if the material in this trans-Neptunian region was too sparse to make big planets, as seems to be the case, there should have been many icy planetesimal-sized bodies present. The gravitational influence of the giant planets, including the outward migration of Uranus and Neptune (Section 1.5), would have depleted the inner regions of this trans-Neptunian population, with many bodies being flung out of the Solar System and others creating the hugely distant Oort cloud. But the inner trans-Neptunian region would not have been emptied, and the more distant regions even less so.

Further evidence of a belt of trans-Neptunian bodies is supplied by the comets (Section 1.2). These are divided into long-period comets and short-period comets, the somewhat arbitrary dividing line being an orbital period of 200 years. In about 1970 it was realised that the rate of discovery of short-period comets was too high for all of them to have an Oort-cloud origin. Such an origin would initially place the comet on a long-period orbit. To become short-period a comet would have to be strongly influenced by the gravity of a giant planet. But the long-period comets had, between them, the full range of orbital inclinations, so the majority could not interact with the giant planets in their low inclination orbits sufficiently often to be converted into short-period comets at the observed rate. Is was realised that a second source of comets was needed.

In 1980, in a paper in *MNRAS*, the Uruguayan astronomer Julio Fernandez speculated that a comet belt from 35 AU and 50 AU was needed to account for the short-period comets. In 1988 the Canadian team of Martin Duncan, Tom Quinn and Scott Tremaine ran computer simulations and found that it was indeed the case that the Oort cloud could not account for the short-period comets, particularly, as noted above, because they occupy rather low inclination orbits, whereas long-period comets have the full range of inclinations.

Other indications that there was once a large population of trans-Neptunian bodies, from which a remnant should have survived, include the following:

- To tilt the rotation axes of Uranus and Neptune to 97.8° and 28.3° respectively, from surely much smaller values at formation, requires each to have suffered a collision with at least one object of around an Earth-mass. The required probability needs roughly 10 bodies of

FIGURE I.2 The Milky Way – our view of the disc of our Galaxy. (Naoyuki Kurita, by permission)

FIGURE 1.7 Eight planets (not to scale). From top to bottom: Mercury, Venus, Earth, Mars, Jupiter, Saturn, Uranus, Neptune. The Moon in top right. No comparable image of Pluto is available. (NASA/JPL-Caltech, PIA 03153)

FIGURE 2.1 Right: A replica of the telescope with which William Herschel discovered Uranus. (Herschel Museum of Astronomy, Bath, UK, by permission)

FIGURE 2.8 The Pluto discovery telescope, a 13-inch diameter astrograph. (Lowell Observatory, by permission)

FIGURE 3.2 The Hubble Space Telescope, launched in 1990. It is a 2.4-metre reflector. (NASA)

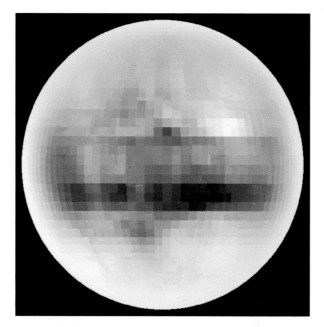

FIGURE 5.7 A two-colour map of the Charon-facing hemisphere of Pluto, derived by Eliot Young and colleagues in 2001, from Charon's transits of Pluto.

FIGURE 8.2 The launch of the Atlas V rocket from Cape Canaveral on 19 January 2006 with the *New Horizons* payload. (NASA)

FIGURE 8.3 The *New Horizons* payload, showing the location of its seven instruments. (NASA)

FIGURE 9.2 An atrist's impression of the view from *New Horizons* at Pluto. The Sun is the brilliant light above the crescent Charon. (NASA/JHUAPL/SwRI)

around an Earth-mass in the Uranus-Neptune region early in Solar System history.

- The capture of Triton (Section 4.3) requires many Triton-like objects in the Neptune region to raise the capture probability. Pluto would be the largest survivor in this region.
- The impact origin of Charon (Section 4.3) is only feasible if lots of potential colliders of 30–50% the mass of Pluto orbited in the region of Pluto.
- The discovery of the small icy body 2060 Chiron by Charles Kowal in 1977 orbiting between Saturn and Uranus, and 5145 Pholus in 1992 in a similar orbit, both orbits unstable on the order of 100 Myr, indicated replenishment from some outer reservoir.

So, where is the outer reservoir, where is the remnant population of small bodies in the region beyond Neptune?

6.2 THE TRICKLE AND THE FLOOD

Tombaugh's search for Pluto had established that there was nothing else beyond Neptune in a moderately low inclination orbit as bright as Pluto. In the 1970s the American astronomer Charles Kowal at the US Naval Observatory, using photography with a blink comparator, searched half the ecliptic zone with the instrumental sensitivity capable of detecting objects nearly 100 times fainter than Pluto, but only found 2060 Chiron, on 01 November 1977, orbiting between Jupiter and Uranus. Its orbit currently has a semimajor axis, $a = 13.708$ AU, an eccentricity, $e = 0.37911$ and an inclination, $i = 6.9311°$. Initially only upper limits could be placed on its diameter, 300–370 km. Present estimates are 142 ± 10 km and 233 ± 14 km. It was the first object to be discovered orbiting (unstably) between Jupiter and Neptune, the first of a new group of icy bodies called Centaurs.

Many Centaurs are now known, all in unstable orbits between Jupiter and Neptune. In Greek mythology the Centaurs are creatures that are human down to the waist and horse below that, an apt name for objects that possess characteristics of asteroids and comets. Chiron is the most important.

But the Centaurs only *indicate* the existence of a population of trans-Neptunian objects; they are not in themselves members of the sought for trans-Neptunian belt of objects.

The discovery that Pluto was not alone beyond Neptune was made by David Jewitt and Jane Luu. They began their search in 1987, when David Jewitt was at MIT with his PhD student Jane Luu. They used telescopes at Kitt Peak in Arizona and at the Cerro-Tololo Inter-American Observatory in Chile, to obtain photographs, pairs of which were compared using a blink comparator. Each pair of plates took about 8 hours to scrutinise. The search was speeded up by the use of CCDs in spite of their then narrower field of view, because 'blinking' could then be performed on a computer screen. Also CCDs collect 90% of incident light, compared with 10% for photographic plates, which is a big increase in sensitivity.

In 1988 Jewitt moved to the Institute of Astronomy in Hawaii, where Luu joined him. They continued their search for trans-Neptunian objects using the University of Hawaii's 2.24-metre telescope on Mauna Kea, using CCDs. After a while they obtained a CCD with 1024×1024 pixels, which increased the field of view and thus speeded up the search. At about this time three other groups started searching. All four groups searched the opposition region, as had Tombaugh in his search for Pluto (Section 2.3).

On 30 August 1992 Jewitt and Luu detected a faint object moving slowly against the stars, over 50 times fainter than Kowal had been able to detect. They obtained its distance from its rate of motion and obtained 37–59 AU from the Earth. Assuming a low albedo, which corresponds to a surface long exposed to radiation, the diameter would be ~260 km. Given an icy-rocky composition this places it on the threshold of being rounded by its own gravity. Their discovery was announced on 14 September 1992. Its provisional name was 1992 QB$_1$.

But it could have been a comet. This possibility was ruled out after several months, by which time its orbit had been established

FIGURE 6.2 David Jewitt and Jane Luu, who discovered, on 30 August 1992, that Pluto was not the only substantial body orbiting beyond Neptune. (David Jewitt and Jane Luu, by permission)

– certainly not cometary, the eccentricity was much too small. The present, not very different, values are $a = 43.7$ AU, $e = 0.0654$, and $i = 2.19°$. It was thus the first of the sought for belt of objects beyond Neptune, called the Edgeworth-Kuiper belt. You will see in Section 6.3 that it belongs to the most populous type of object in the Edgeworth-Kuiper belt, the classical objects, also called the cubewanos ('queue-be-one-owes'), from the pronunciation of QB_1. 1992 QB_1 is about 160 km across. It remains with its provisional name, though it has been given the asteroid number 15760 but, of course, it's not an asteroid.

Searches continued, by Jewitt and Luu and by others, and by late 1994 15 objects in the Edgeworth-Kuiper belt had been discovered, and since 2001 several have been discovered with satellites. By early 2005 a few wide-field surveys had been carried out and 1000 objects were

BOX 6.1 THE PROVISIONAL NAMING OF 1992 QB₁
(PLEASE READ)

The name of a newly discovered small body in the Solar System starts with the year of discovery, in this case 1992. Q denotes the half-month of discovery, with A denoting the first half of January and Y the second half of December (I is not used). Q is thus the second half of August. B_1 shows that the body was the second to be discovered in the second half of August. A few decades ago, the B on its own would have been sufficient to specify that it was second, with A being the first, and so on through the alphabet, omitting I. But in recent years there have been half months in which more than 25 small Solar System bodies have been discovered, so the numerical subscript was introduced to denote how many times the alphabet had been cycled through. In this example, it is the first time.

known, some rivalling Pluto in size. Scaling to the whole ecliptic region gives a few tens of thousands of objects in the Edgeworth-Kuiper belt with diameters larger than about 100 km, and about a hundred larger than 1000 km, all orbiting 30–50 AU from the Sun. Discoveries continue, including further large ones.

Overall, the Edgeworth-Kuiper belt is estimated to contain 20–200 times more mass than does the asteroid belt, the latter having an estimated mass rather less than a thousandth of the mass of the Earth. The present mass in the belt is surely a mere fraction of the mass initially in this region, which simulations of Solar System formation suggest was 15–50 Earth masses.

6.3 KUIPER BELT OBJECTS

First, some abbreviations. The Edgeworth-Kuiper belt I'll abbreviate from now on to the E-K belt. Objects in the E-K belt are sometimes referred to, quite logically, as E-K objects, but the name in widespread use is Kuiper belt objects, abbreviated to KBOs. Fairly widely used is

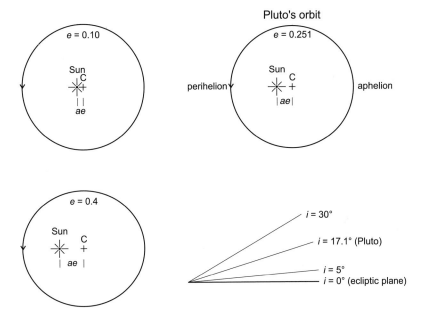

FIGURE 6.3 Illustrative orbits to help you picture the actual orbits of the known KBOs. To aid comparison all three orbits have the same semimajor axis.

the name Kuiper belt for the E-K belt, but I dislike it because it would leave Edgeworth's contribution completely forgotten.

Trans-Neptunian objects (TNOs) include all bodies in orbits with semimajor axes larger than Pluto's, and thus include the Oort cloud objects. This chapter focuses on the E-K belt and its members, the KBOs. In this section the orbits, sizes, masses and compositions of KBOs are discussed.

Orbits of KBOs

Here I will concentrate on the actual orbits, sizes, eccentricities and inclinations; the origin of the various features of the orbits is a subject for Section 6.4.

Figure 6.3 shows four illustrative orbits to help you picture the orbits of the known KBOs. Three eccentricities are shown, low (0.10), moderately high (0.251, Pluto) and high (0.4), and four inclinations, zero, low (5°), moderately high (17.1°, Pluto) and high (30°). Note that even the most eccentric orbit shown is not very elliptical but, due to the corresponding large offset of the centre C from the Sun, the differences between the perihelion and aphelion distances are much larger than when the eccentricity is only 0.1.

If the instantaneous positions of all the known KBOs were shown on a diagram they would be contained within a fuzzy surface rather like a torus (a doughnut, the type with a hole), with an inner boundary beyond Neptune. Three distinct sorts of KBO orbits are recognised.

First, there are the classical KBOs (also called the cubewanos). These account for about two thirds of all known KBOs. They occupy orbits that are no more than moderately eccentric, and so they are confined to the zone 30–55 AU from the Sun. Their semimajor axes range from a little over 40 AU (safely beyond the 3:2 mean motion resonance with Neptune at 39.4 AU), to 47.7 AU. The latter value corresponds to the 2:1 mean motion resonance with Neptune where, for every two orbits of Neptune a body orbits once. There is a sharp decrease in the number of classical KBOs at larger semimajor axes, a feature called the Kuiper cliff. This might be the outer edge of the classical belt or the inner edge of a broad gap – objects have been detected around the 5:2 mean motion resonance at 55 AU. Between 42 AU and 47.7 AU other mean motion resonances with Neptune have left their mark by carving narrow strips of instability.

Classical KBOs are divided into two categories. Those in low eccentricity, low inclination orbits, ranging up to about 0.1 and 10° respectively are called dynamically cold, not because they are at exceptionally low temperature, they're not, but because their speeds relative to each other are low. The rest are in higher eccentricity, higher inclination orbits, ranging up to about 0.2 and 30° respectively. These are dynamically hot because of their higher relative speeds.

Second, there are the plutinos. These occupy orbits in the same 3:2 mean motion resonance with Neptune as is the case for Pluto's orbit (Section 2.5). Most have orbital eccentricities of order of that

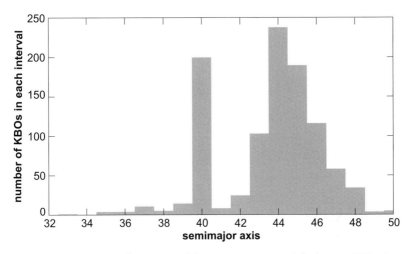

FIGURE 6.4 A histogram of the semimajor axes of the known KBOs, in the range 32–50 AU.

of Pluto, but orbital inclinations ranging from around Pluto's downwards. Many, like Pluto, have perihelia within the orbit of Neptune, but because of the resonance they do not approach that giant planet very closely (Section 2.5). The 200 or so known plutinos constitute about 10% of the known KBOs. Pluto is the largest plutino. Inwards of the resonance at 39.4 AU, the population of KBOs is very sparse.

Figure 6.4 shows a histogram of the semimajor axes of the known KBOs in the range 32–50 AU. You can see the various features that I've described.

Third, there are the scattered disc objects (SDOs). This is a rather sparse population occupying unstable orbits typically with much greater orbital eccentricities and inclinations than found among the classical KBOs and the plutinos, some SDOs having aphelia beyond 1000 AU. Their name derives from the view that they were scattered outwards by Neptune; more on this, and on the origin of the classical KBOs and the plutinos in Section 6.4.

Several known KBOs have satellites, some comprising binary systems, like Pluto and Charon. Recall that the definition of a binary system is that the centre of mass of the pair is not within either body

Table 6.1 *The Kuiper belt objects with diameters exceeding 800 km.*[1]

	Type	Diameter (km)[2]	Mass (10^{22} kg)[3]	n	A (AU)	e	i (°)	Geometric albedo (%)[4]
Eris	SDO	2600 (−200/+400)	1.67 ± 0.02	1	67.67	0.442	44.19	86 ± 7
Pluto	Plutino	2304 ± 20	1.302 ± 0.007	3	39.64	0.251	17.14	55
Makemake	Classical	1500 (−200/+400)	~0.4	0	45.79	0.159	28.96	78 ± 9
Haumea	12:7 res.	1960×1518×996	0.400 ± 0.004	2	43.13	0.195	28.22	~80
Sedna	Detached	1200–1600	0.08–0.7 {small!}	0	525.86	0.855	11.93	16–30
2002 TC$_{302}$	5:2 res.	1150	0.008–0.18	0	55.24	0.292	35.04	~5
Orcus	Plutino	946 ± 73	~0.07 {small!}	1	39.19	0.226	20.59	19.8 ± 3
Varuna	Classical	~900 (uncertain)	~0.04?	0	43.13	0.051	17.2	3.7–26
Quaoar	Classical	840 (−190/+207)	0.10–0.26	0	43.61	0.038	7.99	19.9 (−7/+13)
2003 MW$_{12}$	Classical	~838 (500–1130)	~0.06?	0	45.95	0.138	21.49	9?

[1] n is the number of satellites, a is the semimajor axis of the orbit, e its eccentricity and i its inclination.

[2] The equatorial diameter of the Earth is 12 756 km. KBOs that are not near spherical are given a mean diameter.

[3] The mass of the Earth is 597.42 × 10^{22} kg

[4] 86% implies fresh deep snow (not necessarily H_2O), and 9% is typical of carbon-rich soots.

but in the space between them, as discussed in Section 4.3 for the case of Pluto and Charon (because of their relatively large sizes, the term double planet is also used for Pluto and Charon). It is estimated that at least 5% of KBOs are binary systems.

Sizes of KBOs

The sizes of the known KBOs range from a few thousand kilometres diameter downwards. Table 6.1 lists the names and several details of the ten largest KBOs. Except for Pluto, I have compiled this table by obtaining data on the individual objects from various websites. Do not be surprised if you find somewhat different values in your own searches; some data are rather uncertain. In most cases in Table 6.1 uncertainties in the values of diameter, mass and albedo are given. In the few others an uncertainty around ±10% is reasonable. Some of the diameters, masses and albedos in Table 6.1 are indicative rather than definitive, surely subject to revision. Pluto is an exception, because it is relatively close, relatively large, and has been observed intensively.

The first column in Table 6.1 gives the name of the KBO. The second column lists the type. Haumea is a classical KBO in a 12:7 resonance with Neptune. Sedna has such a large orbit that it is labelled 'detached'; it might even be an inner member of the Oort cloud. You can check the allocated type versus the orbital elements given in columns 6–8.

The diameters (third column) show that even though Pluto was the first KBO to be discovered, and is the largest plutino, it is only the second largest known KBO, the scattered disc object (SDO) Eris being larger. The three largest KBOs, Eris, Pluto and Makemake, are large enough for self-gravity to make them spherical. The fourth largest, Haumea, is elongated; its longest dimension is given, along with the largest and smallest dimensions of a cross-section half way along its length. I'll return to Haumea shortly. The diameters/dimensions have been determined by various methods, for example by

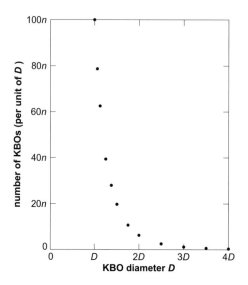

FIGURE 6.5 The number of KBOs versus their diameter D (in arbitrary units). The number is proportional to $1/D^4$, which is broadly in accord with the observational data (though many of the smaller objects await detection).

- direct imaging (for those with angular diameters greater than about 0.03 arcsec, which corresponds to a body with a diameter of around 900 km at a distance of 40 AU)
- measurement of the amount of reflected light, plus an assumed albedo (Section 3.1)
- combined measurements of albedo and thermal emission
- assumed densities (if masses are known from satellite orbits).

What about the size distribution of the *whole* population of KBOs, including those yet to be discovered? To obtain this we must use a model of their formation. It is beyond reasonable doubt that they formed from the dust in the outer solar nebula by accretion (Section 1.5), whereby small particles adhered when they met and built larger particles. When these larger particles met, if the collision was gentle they too stuck together, and so on until bodies were too thinly dispersed to accrete further. Accretion models show that the larger the object the fewer there will be. Figure 6.5 shows this in a quantitative way, as a graph of the number of KBOs versus their diameter, D. In this particular model, the number is proportional to $1/D^4$, corresponding to a very large decrease in the number as D increases. For example a doubling of D corresponds to a decrease in the number by a factor of 16.

What of the measured sizes? Direct imaging is only possible for the largest objects. To obtain values for the others, the second of the four methods listed above has been used. The known KBOs lie fairly close to the trend in Figure 6.5, though many of the smaller objects await detection.

Masses of KBOs

The masses of the KBOs in Table 6.1 are given in column 4. You can see that of the four largest KBOs, Makemake has the least certain mass. You can also see from column 5 that Makemake is the only one of the four without a satellite; recall from Section 1.6 that a small satellite enables the mass of its planet to be determined accurately. Though Orcus has a satellite, the mass of Orcus awaits accurate determination. The others tabulated have no known satellites, have poorly known sizes, and masses that have to be estimated from assumed densities and are consequently poorly known. The majority of KBOs have had no satellite detected as yet.

What about the total mass in the E-K belt? At the end of Section 6.2 you saw that the estimated mass today is 20 to 200 times more than that of the asteroid belt. This is of the order of a tenth of the Earth's mass in the E-K belt. You also saw that the present mass in the belt is a mere fraction of the mass initially in this region, which simulations suggest was 15 to 50 Earth masses. Section 6.4 explores why there could have been so much mass in the E-K belt, and why so little of it remains.

Compositions of KBOs

One indication we have of the compositions of the KBOs is their mean densities, which you will recall is the mass divided by the volume. For KBOs with companions we can measure the masses, and if we have the size we get the density. Unfortunately, for any KBO lacking a companion, as pointed out above, the mass can only be obtained from the size and an assumed density. Unfortunately, this requires an assumed composition. What is a reasonable assumption?

There can be little doubt that the KBOs, small bodies born from outer Solar System materials and dwelling there ever since, are icy-rocky bodies. In Section 3.3 the conclusion was reached that about a third of Pluto's mass consists of icy materials, mainly H_2O, and the rest mainly hydrated rocky materials. In Section 5.4 an interior model was presented.

Whereas the mean density of Pluto is 2030 kg/m^3, that of Eris, using the diameter and mass in Table 6.1, is around 1800 kg/m^3, which suggests a broadly similar composition, though with a slightly greater fraction of icy materials in Eris.

The non-spherical Haumea (shape determined from its lightcurve) has a density somewhere between 2600 and 3300 kg/m^3, indicating a mainly rocky composition. Why is this, and why is such a large body so non-spherical? From its lightcurve we know it to be a fast rotator, with a rotation period of only four hours. You might think that this explains its elongated shape, but such a large rocky body could not be elongated anywhere near enough. And why is it rotating so rapidly? The answer to these three questions is provided by a model in which Haumea collided with another body. This removed its icy materials, distorted its shape, and increased its rotation rate. Support for this model is the detection of what are surely collisional fragments, namely Haumea's two small satellites, and small bodies in broadly similar orbits to Haumea. Note that some icy materials remain – infrared spectra have revealed Haumea to be covered in H_2O ice.

For the remaining KBOs in Table 6.1 useful densities cannot be calculated, nor indeed for any of all the others with no detected satellites, though, apart from oddities like Haumea, there is no reason to think that they are not icy-rocky bodies.

A further indication of KBO compositions has been obtained by analysing the light reflected by them. Initially only the colours of KBOs were measured, showing them to range from grey (neutral), i.e. they reflect sunlight uniformly across the solar spectrum, to red-tinted, i.e. they reflect more of the red end of the solar spectrum. This indicates surface compositions ranging from (dirty) ices to dark hydrocarbons. Astronomers had expected KBOs to be uniformly dark, with the more volatile ices having been converted by cosmic rays into heavier hydrocarbons, which are darker and red-tinted. Why so varied? Resurfacing by impacts was shown by spectral analysis in 2001 to be insufficient to explain the spread in colours. Resurfacing by outgassing has also been suggested, and remains a possibility.

Among the classical KBOs the 'cold' population members are more homogeneous in colour than the members of the 'hot' population. Whereas the 'cold' ones have a red tint, many of the 'hot' ones are more neutral in colour. Another difference is that binaries are more common in the 'cold' population than in the 'hot' population. This suggests that the two populations are from different sources, a possibility explored further in Section 6.4.

Spectrometers, on the whole, are incapable of detecting spectral features on the faint, distant KBOs, but in 1996 Robert H Brown and colleagues detected CH_4 ice on the surface of the 360-km diameter plutino 1993 SC. Its surface spectrum resembles those of Pluto and Triton, but this does not imply that any other plutino has a similar spectrum. H_2O ice has been detected on several KBOs, including the several hundred kilometre diameter classical KBO 1996 TO_{66}, the 500 km diameter plutino Huya and on the classical KBO Varuna (Table 6.1). In 2004 Michael Brown and colleagues detected crystalline H_2O ice and ammonia hydrate on the classical KBO, Quaoar (Table 6.1). Both of these substances would have been destroyed in less than the age of the Solar System, indicating resurfacing either by internal activity or by meteorite impacts. CH_4 and CO have been found on the largest KBOs (including Pluto).

These results from spectrometry confirm the expectation that, as well as Pluto, many KBO surfaces are covered with a variety of icy materials in various conditions.

Surface temperatures of KBOs have been measured by detecting the infrared radiation that they emit. The values are around 50 K, depending on the fraction of the intercepted solar radiation that is reflected back to space. Except perhaps for the largest KBOs, internal heat surely has a negligible effect on surface temperature.

6.4 THE ORIGIN AND EVOLUTION OF THE E-K BELT

From earlier chapters you have seen that the KBOs are thought to have originated from the dust component of the outer region of the solar nebula. Here I'll examine the origin in more detail.

I noted earlier that the total mass in the KBOs today is estimated to be of the order of a tenth of the Earth's mass. Models of Solar System formation have an initial mass beyond Neptune of about 15 to 50 Earth masses in the form of planetesimals and smaller bodies. Such a large mass is required to enable the accretion of KBOs with diameters in excess of about 1000 km in times less than the age of the Solar System. Much must have been cleared by Uranus and Neptune as they migrated outwards (Section 1.5). Consequently, many small bodies would have been flung out of the Solar System, sometimes with the aid of Jupiter and Saturn's gravity. Other bodies did not quite escape but were retained to form, for example, the Oort cloud, where the initially highly eccentric orbits were circularized by the gravity of the Galaxy and by passing stars. Thus was the Oort cloud created.

As in the case of Jupiter and the asteroid belt, orbital perturbations by the giant planets resulted in large relative velocities, replacing growth by disruption, thus preventing the formation of bodies much larger than the largest we see in the E-K belt today.

After Neptune stopped migrating, its gravitational influence was still present, though further material could have been removed from the trans-Neptunian region more effectively by a passing star and/or by the ejection of dust from collisional grinding through various effects of solar radiation on dust-sized particles. There is now an equilibrium between dust creation and removal, with a low concentration of dust beyond Neptune.

That's the overall picture. What about the more detailed sculpting of the trans-Neptunian populations that we see today? There follows a set of plausible scenarios, often based on computer modelling, but I must emphasize that much more understanding is required.

The plutinos

You have seen that the plutinos are defined by having orbital periods in the 3:2 mean motion resonance with Neptune, which means that their semimajor axes are close to 39.4 AU.

It is widely thought that the plutinos started life in smaller orbits than those they now possess. This is also the case for Uranus and Neptune; the nebula thins with increasing distance from the Sun so these two planets need to have formed closer in, where there was enough material in the solar nebula to enable them to grow before nebular gases were swept away (Section 1.5). This happened when the Solar System was rather less than 10 Myr old, a small fraction of its present age of 4600 Myr. So why did Uranus and Neptune migrate outwards? In most models it's due to a variety of subtle gravitational interactions with the remnants of the disc and with the remnant planetesimals. One model puts Neptune's birthplace in an orbit only 11.5 AU from the Sun. In this model Uranus was born *further out*, at 14.2 AU. Subsequent gravitational interactions between all four giant planets soon resulted in Uranus and Neptune swapping positions well before their final orbits were reached.

I emphasise that this is the outcome of one particular computer model. The important, more general, points are that Uranus and Neptune were born much closer to the Sun than their present orbits of 19.2 AU and 30.2 AU respectively, and that extending from beyond their original orbits there was a large population of planetesimals, of various sizes, some large, extending from, for example, 15 AU to

35 AU (though with no sharp cut-off at 35 AU). The outward migration of Neptune, when it had become the outer planet (perhaps it always was), caused its 3:2 resonance to sweep through the planetesimals, resulting in the capture of some of them into the resonance. A 3:2 resonant orbit is stable; a crude picture is that a planetesimal drops into a trough from which escape is difficult. The planetesimals thus move outwards and, when Neptune stops migrating, there they are, constituting the plutinos. It is the outward sweeping by the resonance that has increased the orbital inclinations and eccentricities. The observed eccentricities are mainly in the range 0.07–0.35, spread fairly uniformly between these values; the inclinations are mainly in the range 1–30°, with a concentration towards lower values.

The sparse population of KBOs inward of 39 AU cannot be explained by Neptune's resonances in its present orbit, but by unstable resonant orbits sweeping through this region as Neptune migrated outwards. Much material would have been ejected, but not before Triton was captured by Neptune, which requires many hundreds of objects of the order of Triton's size in the sparse region for this to be sufficiently likely (Section 4.4).

Classical KBOs
You saw above that the classical KBOs have semimajor axes ranging from a little beyond 40 AU (safely beyond the 3:2 orbital resonance with Neptune at 39.4 AU), to the Kuiper cliff at the 2:1 orbital resonance, at 47.7 AU. You also saw that two sub-populations have been identified, a majority that are dynamically 'cold' and a minority that are dynamically 'hot'.

The dynamically cold population, which constitutes about 40% of the classical KBOs, is believed to have formed more or less where it is now, though at least some of these KBOs might have been moved outwards by Neptune's outward migration. The remainder, the dynamically hot population, is believed to have formed near Neptune and then been ejected outwards by the giant planet, the bodies thereby having their inclinations increased.

Possible explanations of the unanticipated Kuiper cliff at 47.7 AU, beyond which there is a paucity of KBOs, include the sparseness of material beyond it, gravitational scattering by an Earth/Mars-sized planet in the E-K belt, and the gravitational influence of a passing star. The first possibility is inconsistent with the likelihood that the solar nebula had no sharp edge at around 50 AU.

Scattered disc objects (SDOs)

You have seen that these have rather wild, unstable orbits with large eccentricities and inclinations. They get their name from models that indicate that they were scattered outwards by Neptune, particularly during its outward migration. Subsequently, it is possible that there has been a steady supply of SDOs scattered from the classical E-K belt and the plutinos.

You have learned that the majority of the classical KBOs were also emplaced by Neptune. The difference could lies in the randomness of the scattering process, some scattered objects acquired stable orbits (contributing to the classical KBO population), the others acquired unstable orbits (the SDOs). However, this would lead to a continuous range of values of orbital eccentricity and inclination through the classical belt and the SDOs, which is not the case. This is another example of the poor state of our understanding of the E-K belt.

KBOs with satellites

Several of the larger KBOs have one or more satellites (see Table 6.1). When the overall population is examined it is found that a higher proportion of the larger KBOs have satellites than do the smaller KBOs, suggesting a different formation mechanism. Whereas collisional formation is the favoured mechanism for forming satellites of the larger KBOs, this does not work as well for the smaller KBOs. At present this is an unsolved problem.

You saw that there are also binaries, e.g. Pluto and Charon. Indeed, extrapolating from the known cases, the proportion of KBOs in binaries is estimated to exceed 5%. Why so large? Whereas the collision on Pluto resulted in the formation of Charon from a massive disc of debris, the binaries with masses considerably smaller than Pluto and Charon cannot have formed through collision. This is because the gravitational attraction between them would have been so small that their collision speed would have been low, and collisional debris would have been too little. One possibility is low speed collisions (<100 m/s) between two KBOs of comparably low mass, resulting in a bounce apart with little disruption. Such a bounce robs the two bodies of relative speed, enabling a binary to form. Another mechanism has a system with many small KBOs, their gravitational interaction resulting in two of them spiralling together to form a binary. A few other mechanisms have also been suggested.

All the mechanisms that produce satellites are too infrequent to work now, which is another indication that a lot more material was present beyond Neptune than there is today.

6.5 CENTAURS AND SHORT-PERIOD COMETS

I conclude this chapter with a short section on bodies that had their origin in the E-K belt but have since moved inwards, namely the Centaurs and the short-period comets.

The great majority of classical KBOs and plutinos occupy orbits that range from being fairly stable to very stable. By contrast, the SDOs occupy significantly less stable orbits. Consequently, there is a steady trickle inwards. Do we see these erstwhile SDOs? The Centaurs are a population of small icy-rocky bodies, orbiting between Jupiter and Neptune. A few hundred are known, but there must be many more. Their orbits are unstable on a time scale of about 100 Myr, about 2% of the age of the Solar System, which indicates replenishment from some outer reservoir. Computer models show that SDOs are the likely source of the Centaurs.

Models also show that the Centaurs' unstable orbits result in the great majority moving inwards, and acquiring eccentric orbits with no more than modest inclinations. As they approach the Sun

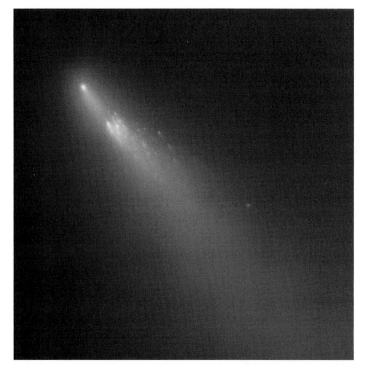

FIGURE 6.6 The short-period comet 73P/Schwassmann-Wachmann 3, which has the short orbital period of 5.3 years. This image, taken by the Hubble Space Telescope in 1995, shows the comet beginning to break up. It is not long for this world! (HST image)

their icy material would begin to vaporize and they would become comets. This is exactly what we observe as the short-period comets (Figure 6.6). The long-period comets come with the full range of inclinations, almost entirely from the Oort cloud.

One new short period comet appears, on average, about every 200 years. To prevent depopulation of the Centaurs, and assuming a steady state Centaur population, SDOs must be converted into Centaurs at the same rate.

Neptune's Trojans

A population that I've not yet mentioned is Neptune's Trojans. These consist of a dozen or so small bodies that cluster around Neptune's orbit, some of them around 60° ahead of Neptune, some 60° behind; they are in a 1:1 mean motion resonance with Neptune. The points on the orbit *exactly* 60° ahead and behind are two of Neptune's five Lagrangian points. The region around each of them offers stable orbits, therefore many of these Trojans probably formed along with Neptune rather than being captured later, though others could have been captured during the outward migration of Uranus and Neptune. There are surely a lot of smaller Trojans awaiting discovery.

Trojans were first named for the 3000 or so small bodies at the corresponding Langrangian points of Jupiter. Trojans were the inhabitants of the ancient city of Troy in present day Turkey.

7   Is Pluto a planet?

What an extraordinary question! Mercury, Venus, the Earth, Mars, Jupiter, Saturn, Uranus and Neptune are planets, so why not Pluto? Pluto's planetary status had been questioned by some astronomers from not long after its discovery, on the basis of its small mass and eccentric, inclined orbit. But the crunch came in 2006. It was in that year that, after much debate and several votes, the International Astronomical Union, at its triennial General Assembly in Prague in August, which I attended, classified Pluto as a *dwarf* planet. This short chapter is devoted to Pluto's classification, which is an ongoing issue. But first let's consider the wider issue of the role of classification in science.

7.1 THE ROLE OF CLASSIFICATION IN SCIENCE

In science, classification provides an economy of description, a tool for structuring knowledge, and can also lead to deeper understanding.

A simple example is provided by crystals. All crystals share two attributes that define the class:

- the basic unit, be it an atom or a molecule, is arranged in one of a variety of repeating patterns in space
- they are solids, i.e. they retain their external form and do not flow like liquids.

The economy of description is that, in place of saying 'one form of water ice is a solid with its component molecules arranged in a repeating pattern in space', one just says 'crystalline water'. Classes are usually defined by more than two attributes, so this economy of description is much more marked than in my example of crystals.

The structuring of knowledge is that when I refer to crystals, others will know that I'm referring to that class of solids in each of which the basic component is arranged in one of a variety of repeating patterns in space.

An example of deeper understanding is provided by the fact that crystals are not all identical. By studying all the different types of crystal we can improve our understanding, for example, of how crystals form. This, in turn, can lead us to a better understanding of the forces between atoms and molecules.

Before the atomic structure of crystals was established, just before the First World War, by a technique called X-ray diffraction, crystals had to be classified by their external form. Interpretation of the results from X-rays was aided by this external classification, and therefore is another example of classification leading to deeper understanding.

A class necessarily divides a set of objects from all other objects, depending on the attributes that define the class. Thus, the crystal class excludes all amorphous solids, i.e. solids with no repeating pattern in space, such as glass, and all liquids and gases. It also excludes complex structures like a stick of wood. With a class being defined on the basis of one or more common attributes, the members of a class will usually differ in other attributes. This leads to the recognition of sub-classes. In the case of crystals, two examples of sub-classes are illustrated in Figure 7.1, where the basic units shown stack in all directions until the surface of the crystal is reached. In the one case the atoms are arranged in a cubic form, and in the other in a hexagonal form. As long as you're some distance from the surface, the view around you is independent of which unit you're sitting in. Note that the dots represent the centres of atoms and not the atoms themselves, which fill the space.

Though classification is an essential feature in all branches of science, there are always hard cases at the boundaries. Indeed the boundary of a class is rarely sharp. To continue with the crystal example, no crystal is perfect. They all contain flaws or impurities that

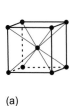 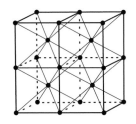 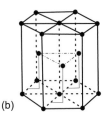

(a) (b)

FIGURE 7.1 Two of the several types of basic form of crystals: (a) cubic (body centred) and (b) hexagonal (close packed). The basic unit shown here is stacked in all directions until the surface of the crystal is reached. In (a) four units are shown stacked together. Note that the dots represent the centres of atoms.

disrupt the regular repeating basic unit. There will be cases where the regular structure is disrupted on a scale of only a few basic units. Is it, overall, a crystal, or amorphous? It's neither, it's somewhere between.

Scientists should not lose sleep over the hard cases. As long as a classification system is beneficial to economy of description, to structuring knowledge and to our understanding, and hard cases constitute a small minority, then keep it. If the system becomes less than useful, then scrap it and replace it with a system based on different shared characteristics.

In the case of planetary bodies, as well as economy of description, useful classifications lead to a better understanding of, for example, the origin of a class, and hence to a deeper understanding of the origin and evolution of the Solar System, which, in turn, will increase our understanding of the many planetary systems being discovered beyond our own.

7.2 PLANETS BEFORE PLUTO'S DISCOVERY: ANTIQUITY TO 1930

When astronomers use the word 'planet' they mean to conjure up a picture of a body with a specific set of attributes. These attributes have changed several times over recorded history.

In antiquity a planet was defined as a celestial body that wandered against the fixed pattern of the stellar background. Indeed, the word 'planet' is derived from the Greek word *planetes*, which means 'wanderer'. Seven planets were visible to the unaided eye. In the English-speaking world these are the Sun (from the old English Sunne), the Moon (from the old English Mona), Mercury, Venus, Mars, Jupiter and Saturn (the names of Roman gods). The Earth was not included because of the prevailing view in antiquity that the Earth was fixed at the centre of the Universe, and not in motion among the stars.

In the early decades of the sixteenth century the Polish astronomer Nicolaus Copernicus (the Latinized version of Coppernic, 1473–1543) revived an ancient Greek idea that the Sun, not the Earth, is fixed at the centre of the Universe. Such a heliocentric Universe had been proposed as early as about 200 BC by Aristarchus of Samos (an island off the coast of what is now Turkey). However, it did not survive long under the weight of Aristotle's influence. The widespread adoption of a heliocentric Universe came after the observations of the Danish nobleman Tycho Ottesen Brahe (1546–1601) and the subsequent analyses of the German mathematician and astronomer Johannes Kepler (1571–1630), which led to Kepler's laws of planetary motion (Section 1.2). These are much more easily framed in a heliocentric system than in any other. The heliocentric Universe at once led to the exclusion of the Sun from the class 'planet', which then comprised Mercury, Venus, the Earth, the Moon, Mars, Jupiter and Saturn.

The discovery in the seventeenth century of satellites orbiting Jupiter and Saturn resulted in a new definition of a planet, well established by the mid eighteenth century – a planet was an object that orbited the Sun. The satellites, including the Moon, though they are carried around the Sun by their planet, orbit their planet rather than occupying independent orbits around the Sun, and were called planetary satellites. Thus, with the Moon no longer regarded as a planet, the number of planets had fallen to six – Mercury, Venus, the Earth, Mars, Jupiter and Saturn.

By the mid-ejghteenth century it was known that comets also orbited the Sun. However, they were not regarded as planets, presumably because, with their huge tails and eccentric orbits, they were so

FIGURE 7.2 Monument in Prague to Tycho Brahe (left) and Johannes Kepler (right).

markedly different in appearance and motion from Mercury, Venus, the Earth, Mars, Jupiter and Saturn.

The number of planets was restored to seven by the discovery of Uranus by Herschel in 1781 (Section 2.1). This was followed in 1801 by the discovery of Ceres, a small body in an orbit not far from where the Titius-Bode rule (Section 2.2) predicted that there should be a planetary orbit. It became the eighth planet. A planet could then be defined as a body that orbits the Sun and obeys the Titius-Bode rule.

Alas! additional small objects were found in the neighbourhood of Ceres, and when the fifth such body, Astrea, was discovered in December 1845, it was soon decided that the five were not planets but minor planets, or asteroids, presumably distinguished from planets on the basis of their size – the largest asteroid, Ceres, was known to have a diameter around 1000 km, whereas the smallest planet, Mercury, was known to have a diameter approaching 5000 km. Within a few decades more asteroids were found, and the great majority were in orbits nowhere near the Titius-Bode orbit. The number of planets had fallen back to seven.

Neptune, discovered in 1846 (Section 2.2), became truly the eighth planet. In the subsequent decades it was thought that a planet occupied an orbit closer to the Sun than Mercury. It was called Vulcan. Its existence was based on an unexplained feature of Mercury's orbit, namely, that its orbit rotated slowly in its orbital plane, the so-called precession of the perihelion. In 1915 Albert Einstein (1879–1955) showed that his General Theory of Relativity, which is an improvement on Newton's Law of gravity, explained fully the precession, thus relegating Vulcan to the dustbin of history. Relegated earlier, in 1879, was a putative planet beyond Neptune, called Hind's planet – earlier claims that it had been seen were discredited.

Then, of course, in 1930, an actual planet beyond Neptune was discovered, Pluto (Section 2.3), in an orbit not in accord with the Titius-Bode rule. Nevertheless there were now nine planets – Mercury, Venus, the Earth, Mars, Jupiter, Saturn, Uranus, Neptune and Pluto. But though these very different bodies were all called planets, and none of the asteroids or comets were called planets, there was no formal definition of a planet at that time.

7.3 THE CLASSIFICATION OF PLUTO

In Section 3.1 you saw that the first estimates of Pluto's diameter were that it was no more than about the diameter of the Earth, and that it could be a lot less. This led some astronomers to speculate that Pluto might be more like an asteroid, or perhaps a large comet. In Section 3.2 you saw that early estimates of Pluto's mass ranged from about the Earth's mass to about a tenth of that value. Generally, subsequent estimates gave even smaller values and by 1978 it was clear that it was far smaller than the Earth. In that year Pluto's satellite was discovered, that enabled Pluto's mass to be determined as about 0.002 that of the Earth, later revised to 0.00218. In the 1990s the first accurate diameter was obtained, 0.18 times that of the Earth, and 0.47 times that of Mercury, though over twice that of the largest asteroid Ceres. The subsequent discovery of KBOs, one of them, Eris, being somewhat larger than Pluto (Section 6.3), has placed Pluto firmly in the class KBOs, and it is the largest known object in the sub-class plutinos. But is it *also* a planet?

In 2005 the US astronomer David Jewitt was of the opinion that, in trying to understand the origin and significance of Pluto, it was better to regard it as a typical KBO rather than as a peculiar planet with a very small mass and odd orbit. Astronomers were divided in their opinion. Surely it's the case that a body can be in two classes, with a different set of attributes placing it in each one? True, but his begs the question of what the attributes are that define the class Planets.

At the end of the twentieth century most astronomers would have agreed with the following three attributes.

1 The object must be in orbit around a star.

   This excludes satellites because their primary orbit is around their planet. It also excludes planetary bodies in interstellar space, which have an uncertain origin, and perhaps were never in orbit around a star.

2 The object must never have generated energy through nuclear fusion.

   This places an *upper* limit on the mass, because larger masses result in core temperatures high enough for nuclear fusion to occur. The threshold is at about 0.08 times the mass of the Sun, or about 80 times the mass of Jupiter, or about 25 000 times the mass of the

Earth. Objects immediately above this mass limit are faint stars with surfaces much cooler than that of the Sun, covering the approximate temperature range 750–2200 K rather than the Sun's 5780 K. They are called brown dwarfs, though they are actually a very dull red or magenta.

3 The object must be large enough for its self-gravity to make it spherical.

The self-gravitational forces inside a body result in an increase of pressure with depth. If the pressure exceeds the strength of the solid materials that constitute the interior, then the materials yield, and a roughly spherical body results, flattened by any rotation, or perhaps elongated by tides due to the proximity of another body. The internal pressures decrease as the size of the body decreases, and there comes a point where the materials do not yield, and so the body need not be even approximately spherical. For a body made of silicates and iron-nickel, the critical diameter is calculated to be about 600 km. These materials have high mechanical strength. Icy solids are less strong, so can be made near-enough spherical at smaller sizes. The known solid bodies in the Solar System confirm, in broad terms, these expectations – see Table 7.1.

As well as the other eight planets, Pluto possesses all three attributes, as does the asteroid Ceres, and several of the larger known KBOs in addition to Pluto. Therefore there must be over ten planets in the Solar System.

Several different additional/alternative attributes were introduced in the early years of this century in an attempt to limit the number of planets in the Solar System, mainly to exclude Ceres and the larger known KBOs other than Pluto, so that the class Planet would still have just the same nine members as before. David A Weintraub (*Further reading and other resources*) gives a very full account of these schemes, none of which was successful. I pick up the story up in August 2006, at the triennial General Assembly of the International Astronomical Union.

Table 7.1 *Sizes and shapes of some smaller bodies in the Solar System.*

Body	Class	Composition[1]	Diameter (km)[2]	Spherical?
Ceres	Asteroid	Rocky-iron	957	Yes[3]
Pallas	Asteroid	Rocky-iron	582×556×500	Not quite
Vesta	Asteroid	Rocky-iron	578×560×458	Not quite
Hygiea	Asteroid	Rocky-iron	500×385×350	No
Enceladus	Saturn satellite	Icy-rocky	513×503×497	Near enough
Proteus	Neptune satellite	Icy-rocky	440×416×404	Not quite

[1] The major components in rocky-iron bodies are silicates and Fe-Ni, or other iron-rich compounds; in icy-rocky bodies the major components are icy materials (mainly water) and silicates.

[2] If the body is 'not quite' or 'almost' spherical, then three dimensions are given, called the principal dimensions. These are the dimensions of what is called an ellipsoid, adjusted to fit as closely as possible the surface of the body. An ellipsoid is characterized by three principle dimensions at right angles to each other, the largest being approximately the maximum dimension across the body.

[3] Ceres rotates once every 9.07 hours. This rather rapid rate has rotationally flattened Ceres so that the polar diameter is 50–60 km less than the tabulated equatorial diameter.

The IAU General Assembly, Prague 2006

The International Astronomical Union was founded in 1919, with a mission to promote and safeguard the science of astronomy in all its aspects through international cooperation. It is acknowledged as the appropriate body to name/number or approve the names of planets, satellites, asteroids, comets, stars and other celestial bodies. In 2005 it set up a seven person Planetary Definition Committee, chaired by the historian of astronomy, Owen Gingerich of Harvard. They brought their definition to the 2006 General Assembly.

At each of several debates open to all of the IAU members at the Assembly, debates which were the subject of huge media coverage, the definition was repeatedly modified. The attributes (criteria) finally adopted by a simple majority of the 424 members present at the final debate, also attracted huge media coverage. The final resolution is as follows ('category' is used in place of 'class').

IAU Resolution: definition of a 'planet' in the Solar System
Contemporary observations are changing our understanding of planetary systems, and it is important that our nomenclature for objects reflects our current understanding. This applies, in particular, to the designation 'planets'. The word 'planet' originally described 'wanderers' that were known only as moving lights in the sky. Recent discoveries lead us to create a new definition, which we can make using currently available scientific information.

Resolution 5A
The IAU therefore resolves that planets and other bodies in our Solar System, except satellites, be defined into three distinct categories in the following way:

1. A 'planet'* is a celestial body that
   a. is in orbit around the Sun,
   b. has sufficient mass for its self-gravity to overcome rigid body forces so that it assumes a hydrostatic equilibrium (nearly round) shape, and
   c. has cleared the neighbourhood around its orbit.
2. A 'dwarf planet' is a celestial body that
   a. is in orbit around the Sun
   b. has sufficient mass for its self-gravity to overcome rigid body forces so that it assumes a hydrostatic equilibrium (nearly round) shape,**
   c. has not cleared the neighbourhood around its orbit, and
   d. is not a satellite.

3. All other objects,[†] except satellites, orbiting the Sun shall be referred to collectively as 'small Solar System bodies'.

    [*] The eight planets are: Mercury, Venus, Earth, Mars, Jupiter, Saturn, Uranus and Neptune.
    [**] An IAU process will be established to assign borderline objects into either dwarf planet or other categories.
    [†] These currently include most of the Solar System asteroids, most trans-Neptunian objects (TNOs), comets and other smaller bodies.

IAU Resolution: Pluto

Resolution 6A
The IAU further resolves:

    Pluto is a 'dwarf planet' by the above definition and is recognised as the prototype of a new category of trans-Neptunian objects.

    The members present at the 2006 General Assembly also agreed that as part of Resolution 6A the IAU should implement a process to establish a name for the new category of large trans-Neptunian objects, i.e. objects that orbit beyond Neptune (which include the KBOs).

    Presumably Pluto was placed in the category dwarf planet because of the other plutinos, which mean that Pluto has not cleared the neighbourhood around its orbit. Other dwarf planets currently recognised by the IAU are the largest asteroid Ceres, and the other three of the four largest KBOs – Eris, Makemake and Haumea, presumably because none of them meets the orbital clearance criterion either. The inclusion of Haumea, which is distinctly non-spherical (Table 6.1) is presumably because we think it has been made non-spherical through a collision (Section 6.3).

The aftermath of Prague 2006

With only a very small fraction of the near 10 000 members of the IAU present for the planet definition debates and votes, it was not surprising that, as well as those present voting *against* the resolutions, there were many absent astronomers who did not like them, with reactions ranging from mild disapproval to fierce hostility.

Among the many objections raised, the following were amongst those most often voiced:

1 If a planet's orbit is regularly crossed by the orbit of another body, then it has not cleared the neighbourhood of its orbit, and therefore Resolution 5A.2c means it would be classified as a dwarf planet. But Pluto, and many plutinos, regularly cross Neptune's orbit. Is Neptune then to be regarded as a dwarf planet? Clearly not! Neptune also has a handful of Trojans in its orbit, and Jupiter has many hundreds (Section 6.5); even Mars has a few. These are also failures to clear the neighbourhood of the orbit, but none of these bodies is a dwarf planet. On the other hand, the chance of an encounter of these objects with Neptune, Jupiter and Mars is pretty well zero. For example, the plutinos are in stable 3:2 resonant orbits with Neptune. Had this criterion been added to the IAU's resolutions then would Resolution 5A.2c been less problematic? Perhaps.

2 Whereas the lower size limit to be a planet is specified, the upper limit is not. An obvious upper limit is that at which nuclear fusion occurs. You have seen that this occurs at masses greater than about 80 times the mass of Jupiter – more massive objects are the brown dwarfs. Brown dwarfs are spherical and some are known orbiting more massive stars in cleared orbits. An upper limit is therefore essential.

3 The meaning of several words/phrases in the resolutions is unclear, as you have seen with 'cleared the neighbourhood around its orbit'.

To these three can be added that any definition of a planet and other bodies in the Solar System should apply to the planetary systems around other stars, the exoplanetary systems, of which over 400 are known, a number that is steadily growing. This would not exclude defining sub-classes containing bodies unlike those found in our System.

Just a week after the Prague resolutions were passed, 353 astronomers and planetary scientists petitioned the IAU as follows:

'We, as planetary scientists and astronomers, do not agree with the IAU's definition of a planet, nor will we use it. A better definition is needed.'

According to the IAU's bylaws, the next opportunity for reconsideration would be at the next General Assembly, in Rio de Janeiro, Brazil, August 2009.

But the Executive Committee of the IAU did not let matters lie, and on 11 June 2008 pronounced in a press release that 'the International Astronomical Union has decided on the term 'plutoid' as the name for dwarf planets like Pluto'. By 'like Pluto' it is meant that the orbit is larger than Neptune's orbit, thus creating a sub-category of dwarf planet that excludes Ceres. This pronouncement, which came as a surprise to most planetary astronomers, has not been well received. My own view is that further debate is needed to establish which, if any, sub-categories of dwarf planet need to be defined and, if so, precise criteria for each one of them need to be specified.

You can see that Pluto is a hard case, along with all the other dwarf planets. There is no dramatic distinction in terms of size. The largest dwarf planet, Eris, has a diameter 0.53 times that of Mercury, which itself has a diameter 0.38 times that of the Earth, and 0.034 times that of Jupiter. Pluto is certainly a KBO and certainly the largest known plutino. Perhaps the category 'planet' could include anything orbiting the Sun with sufficient mass to be spherical. This would increase the number of planets in the Solar System to 13, with the likelihood of further increase as large KBOs are discovered and other large ones already known are classified. It seems to be this growth in numbers that some astronomers have an aversion to.

It must be stressed that it was a vocal minority of astronomers that objected strongly to the 'demotion' of Pluto in 2006, and to the creation of the dwarf planet class. Most planetary astronomers are happy with the creation of the dwarf-planet class.

I dislike the creation of the class 'dwarf planet' separate from the distinct class 'planet'. I would much prefer a class 'planet' with a sub-class 'dwarf planet', and another sub-class, possibly named 'large planet'. But that's another story.

Returning to things as they are, did the IAU General Assembly in 2009 sort things out?

The IAU General Assembly, Rio de Janeiro 2009

No they didn't! The issue was brought up by a group of amateur astronomers early in the meeting, but any discussion was quashed. So, for now, we have to make do with a flawed, incomplete classification system for planets.

With things as they are, is the Solar System 'stuck' with just having eight planets? Not necessarily. Who knows what lurks in the outer depths of the Solar System?

8 The *New Horizons* mission to Pluto (and beyond)

We would clearly learn a lot more about Pluto, its three satellites, the E-K belt, and anything else beyond Pluto, were a spacecraft targeted to investigate this far flung region of the Solar System. As yet no such spacecraft has visited Pluto and beyond, but one is on its way, *New Horizons*.

8.1 THE LONG PATH TO *NEW HORIZONS*

Table 8.1 lists the spacecraft that have already visited the outer Solar System, and reached their targets. You can see that Pluto is the only one of the original nine planets that has not been visited by a spacecraft. Why? One reason is surely that when each of the missions in Table 8.1 was being selected for development, no other KBOs were known, the first to be discovered was the small body 1992 QB_1 in 1992 (Section 6.2), and therefore Pluto and its satellite Charon was regarded as just a small, isolated system. Consequently it was of considerably less interest than it is now, with its numerous companions in the E-K belt. Attention was instead focused on the rich domain of the four terrestrial planets and the four giant planets plus their numerous satellites.

As our knowledge of Pluto grew, so did interest in sending a spacecraft there, such that in the late 1980s a small number of planetary astronomers began to campaign for a mission to Pluto. The campaign was aided by the 1989 flyby of Neptune by *Voyager 2*. In particular, Neptune's largest satellite Triton, with a diameter just a little larger than Pluto's, was found to have a rather similar surface and atmospheric composition to Pluto. Moreover, active geysers on Triton were observed, probably the result of the sub-surface sublimation of nitrogen. Could such geysers also be present on Pluto? Adding

Table 8.1 *Spacecraft missions to the outer Solar System.*[1]

Spacecraft	Mission	Encounter date	Present location[2]	Fate
Pioneer 10	Jupiter flyby	03 Dec. 1973	2003, 82 AU from Sun	Interstellar space
Pioneer 11	Jupiter flyby	02 Dec. 1974	1995, 43 AU from Sun	Interstellar space
	Saturn flyby	01 Sept. 1979		
Voyager 1	Jupiter flyby	05 March 1979	109 AU from the Sun	Interstellar space
	Saturn flyby	12 Nov. 1980		
Voyager 2	Jupiter flyby	09 July 1979	88 AU from the Sun	Interstellar space
	Saturn flyby	25 Aug. 1981		
	Uranus flyby	24 Jan. 1986		
	Neptune flyby	24 Aug. 1989		
Galileo	Jupiter orbiter & probe	07 Dec. 1995	Jupiter orbit Burnt up in Jupiter	Jupiter orbit Vaporized
Cassini-Huygens	Saturn orbiter & Titan lander[3]	01 July 2004	Saturn orbit Titan surface	Saturn orbit Titan surface

[1] All are NASA missions except *Cassini-Huygens*, which is a NASA-ESA mission.

[2] The year given is for the last radio contact received, and the corresponding distance. None of those en route to interstellar space has passed anywhere near Pluto and the other KBOs.

[3] *Cassini-Huygens* flew by Jupiter on 30 December 2000, distantly, at 9.7 million km.

FIGURE 8.1 Alan Stern and Frances Bagenal. Stern's visit to NASA headquarters in 1989 was instrumental in the ultimate approval by NASA of the *New Horizons* spacecraft, now en route to Pluto and beyond. (Alan Stern and Frances Bagenal, by permission)

to the growing interest in Pluto was the view that Triton was captured by Neptune and that before capture it was a freely orbiting twin of Pluto (Section 4.3).

Among those pressing NASA for a mission to Pluto were The US astronomers Frances Bagenal and Alan Stern. Stern's visit to NASA headquarters in 1989 was instrumental in the positive outcome, and studies for a low-mass (350 kg) Pluto mission began that year. In spite of scepticism among many in NASA, the studies continued. The need for a mission was urgent, for three reasons.

First, after Pluto's perihelion in 1989 its distance from the Sun began to grow, making the journey to Pluto longer, and weakening the data-carrying radio signals that would be received on Earth.

Second, after perihelion the solar insolation (the solar radiation falling on Pluto) began to fall, which must ultimately result in the precipitation of nearly all of its atmosphere on to the surface, changing the surface as well as the atmosphere, and leaving little atmosphere to study (Section 5.3). Pluto's perihelion and aphelion distances are, respectively, 29.7 AU and 49.6 AU, corresponding to a decrease in insolation of $(29.7/49.6)^2$, which is 0.36, and so at aphelion the insolation is only 36% that at perihelion. For comparison, the insolation on Mars is about 43% that on the Earth, not much greater than the variation experienced by Pluto in its orbit. It was uncertain when, in the next few decades, Pluto's atmosphere would begin to precipitate, how quickly, and to what extent. Aphelion is not until 2113, but the rate at which Pluto recedes is a maximum around perihelion (Kepler's second law, Section 1.2). Pluto needed to be visited before much of its atmosphere had frozen.

Third, on Pluto, winter in its southern hemisphere was fast approaching, and with Pluto's large axial inclination of 57.5° (retrograde) with respect to its orbital plane, an increasingly large proportion of its southern hemisphere was being plunged into darkness: by 2015 almost half will be too dark for much to be discerned. The same is the case for Charon. The sooner the arrival of a spacecraft, the more of the globe would be sunlit at some time during Pluto's and Charon's day.

The first study of a Pluto mission was in 1990, by the Discovery Working Group led by Robert Farquhar (NASA Goddard Space Flight Center). Arguments brought to NASA yielded fruit in 1991 when NASA's Solar System Exploration Subcommittee placed a Pluto flyby in the highest priority category for new missions in the 1990s. Much discussion followed, with various proposals to NASA for spacecraft and launch vehicle coming and going (see Stern and Mitton in *Further reading and other resources* for details). After all this, in the autumn of 2000 NASA's administrator for Space Science, Edward Weiler, announced that there would, after all, be no mission to Pluto. This led to a storm of protest, not just from 'plutophiles' and other planetary astronomers, but from the broader public. There was much media coverage, overwhelmingly in favour of a Pluto mission.

Within a few weeks, in the week before Christmas 2000 Weiler invited proposals to NASA for a low budget mission ($500 million) to Pluto *and the Edgeworth-Kuiper belt* (a previous proposal for a Pluto-E-K belt mission had been unsuccessful). The deadline was March 2001, then put back to April. Five proposals were submitted. A short-list of two was announced in June 2001, then in November 2001 NASA announced the selection of the *New Horizons* proposal, from Alan Stern *et al.*, as being the better of the two for performing science at Pluto and in the E-K belt, and the more likely of the two to meet the launch target of January 2006, and do so within the budget.

Alas! In early 2002 the budget submitted to the US Congress by the US Office of Management and Budget showed *New Horizons* cancelled, on the basis of cost estimates that were for an earlier, expensive JPL mission that had already been cancelled! After pressure, again from astronomers and the public, Congress, late in 2002, sent a budget to the US President with *New Horizons* re-instated. The next NASA budget in February 2003 showed *New Horizons* to be fully funded, with a launch date in January 2006. On 19 January of that year the spacecraft was launched successfully from Cape Canaveral in Florida on an Atlas V rocket with a third stage added to increase the speed (Figure 8.2). Alan Stern is the principal scientific investigator.

8.2 MISSION OBJECTIVES

Experience with all the many spacecraft that have visited planets and satellites puts it beyond reasonable doubt that *New Horizons* will yield a cornucopia of new data on Pluto, Charon and other KBOs.

The mission objectives fall into three categories: primary, secondary and tertiary.

The primary objectives are as follows:

- Characterize the global geology and morphology of Pluto and Charon
- Map the chemical compositions of the surfaces of Pluto and Charon
- Characterize the neutral (non-ionized) atmosphere of Pluto, and the atmosphere's escape rate.

FIGURE 8.2 The launch of the Atlas V rocket from Cape Canaveral on 19 January 2006 with the *New Horizons* payload. (NASA) (See plate section for colour version.)

Failure to achieve any of these would be regarded as failure of the mission.

Here are the secondary objectives:

- Characterize the time-variability of Pluto's surface and atmosphere
- Obtain stereo images of selected areas of Pluto and Charon
- Map the terminators (day-night border) of Pluto and Charon at high resolution
- Map the chemical compositions of selected areas of Pluto and Charon at high resolution
- Characterize Pluto's ionosphere, and its interaction with the solar wind. (An ionosphere is the outer region of a planet's atmosphere, where an appreciable proportion of the gas atoms is ionized, i.e. have had electrons driven off. The solar wind is a variable outflow from the Sun, consisting almost entirely of electrons and protons resulting from the thermal dissociation of hydrogen atoms.)

- Search for neutral molecules in the atmosphere, such as molecular hydrogen H_2, hydrocarbons (compounds of carbon and hydrogen, such as CH_4), hydrogen cyanide (HCN), and other molecules with CN in them (nitriles)
- Search for any Charon atmosphere
- Determine the Bond albedos for Pluto and Charon (the Bond albedo determines the fraction of the intercepted solar radiation that is reflected back to space by the surface and any atmosphere)
- Map surface temperatures of Pluto and Charon
- Explore Nix and Hydra
- Conduct education and public outreach observations of special aesthetic merit.

It would be disappointing if most of these objectives were not met.

The tertiary objectives are as follows:

- Characterize the energetic particle environment of Pluto and Charon
- Refine bulk parameters (radii, masses) and the orbits of Pluto and Charon
- Search for additional satellites, and any planetary rings.

Any/all of these could be dropped, if need be, in favour of the primary and secondary objectives.

Note that the objectives do not include measurement of Pluto's magnetic field. A magnetometer would have added too much mass, and two other instruments, SWAP and PEPSSI (Section 8.2) could indirectly detect any intrinsic magnetic field around Pluto.

The most important questions that *New Horizons* should answer about the Pluto system, or at least go some way towards answering, are as follows:

- Are Pluto and Triton really similar, indicating that Triton was captured from the region of Pluto, or is the similarity superficial?
- Do Pluto's surface markings indicate substantial regional variations in surface composition, or are they the result of variations in the minor components, CO and CH_4 ices, with N_2 ice dominant in most regions?

- What is the internal structure of Pluto? Is there any activity that could, for example, produce geysers as on Triton?
- Are there haze layers in the atmosphere? Are there sharp changes in the atmospheric temperature versus altitude?
- What effects do the large changes in solar insolation, resulting from the orbit's large eccentricity, have on Pluto's atmosphere and surface?
- Is Charon really a result of the impact of a large object on Pluto?

Comparable questions about Charon's atmosphere, surface, and interior could also be answered, in whole or in part, by *New Horizons*.

Other KBOs to be flown by remain to be selected; they'll need to be close to the flight path because of *New Horizons'* limited manoeuvring capability. But along with new data on Pluto's system, new data on the KBOs will certainly improve considerably our understanding of the outer Solar System and its early evolution.

8.3 THE SPACECRAFT: INSTRUMENTATION AND JOURNEY

Instrumentation

The spacecraft at launch had a mass of 478 kg, which includes everything: rocket motors, manoeuvring fuel, and so on, and of course the payload of scientific instruments (Figure 8.3). The payload has a very low mass, a little less than 30 kg and consumes only 28 watts of electrical power, like a dim electric light bulb. The instruments are as follows:

- Alice: an ultraviolet imaging spectrometer that will examine the structure and composition of Pluto's atmosphere and any atmosphere on Charon
- Ralph: a visible and near infrared camera that will obtain high resolution colour maps, surface composition maps and temperature maps of the surfaces of Pluto and Charon (Ralph and Alice were the main characters in the US 1950s television series *The Honeymooners.*)
- Long Range Reconnaissance Imager (LORRI): an optical telescope that will image features on Pluto as small as about 50 metres and, three days after flyby, will image areas not visible at closest approach

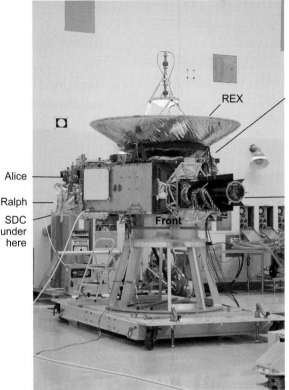

FIGURE 8.3 The *New Horizons* payload, showing the location of its seven instruments. (NASA) (See plate section for colour version.)

- Solar Wind Around Pluto (SWAP): a device that will detect charged particles from the solar wind near Pluto, to determine whether it has a magnetosphere and how fast its atmosphere is escaping (A magnetosphere is the field around the planet produced by its internal magnetic field.)
- Particle Energetic Particle Spectrometer Science Investigation (PEPSSI): a device that would detect any neutral atoms escaping from Pluto's atmosphere and subsequently becoming charged by the solar wind
- Student Dust Counter (SDC): a device that counts and measures the masses of dust particles along the spacecraft's entire trajectory, even beyond Pluto (This includes regions of interplanetary space not yet

sampled, and is the first planetary science spacecraft instrument to have been built by students.)

- Radio Science Experiment (REX, not RSE!): a circuit board integrated with the spacecraft radio-telecommunications system that will obtain data on Pluto's atmospheric structure and surface thermal properties, and enable measurements of the mass of Pluto, Charon and the KBOs encountered.

The instruments are designed to return data to Earth from at least as far away as 50 AU, which is just beyond the classical E-K belt.

The journey

New Horizons' launch on 19 January 2006 on a powerful Atlas V rocket saw it leaving the Earth faster than any other spacecraft, at about 16 km/s. After two small corrections to its trajectory towards the end of the month, and another on 9 March, the spacecraft made a distant flyby, at a range of 101 867 km, of the 2.3 km diameter asteroid 132524 APL. The spectra obtained showed it to be an S-type, i.e. a mixture of Fe-Ni and silicates. In February 2007 the spacecraft interacted gravitationally with Jupiter, on a trajectory such that a tiny fraction of Jupiter's orbital energy was transferred to New Horizons. This raised its speed from much less than 16 km/s (as it had slowed down in its journey away from the Earth and the Sun) to about 23 km/s. Closest approach was 2.3 million kilometres, from where Jupiter extended 3.56° across the sky (our Moon extends just 0.5° across our sky). This planned encounter accelerated the craft significantly, shortening the travel time to Pluto by nearly four years.

During the Jupiter encounter various studies were made of the planet itself, its rings and some of its satellites. A large amount of data was obtained.

Figure 8.4 shows the trajectory of New Horizons, whose plane is inclined at only a couple degrees with respect to the ecliptic plane. New Horizons' position in its trajectory is continuously updated at http://www.yaohua2000.org/cgi-bin/New%20Horizons.pl. On

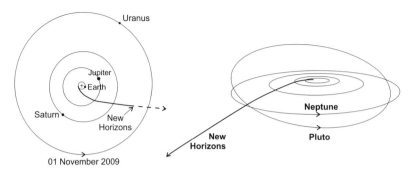

FIGURE 8.4 The trajectory of *New Horizons*. Left: its location and the locations of Jupiter, Saturn and Uranus on 1 November 2009. Right: *New Horizons'* path into the E-K belt.

FIGURE 8.5 *New Horizons nears* Pluto (lower left), with Charon, Nix and Hydra in view – an artist's impression. (NASA/JHUAPL/SwRI, PVSS00039)

1 November 2009, at 17:04 UT it was 14.90139 AU from the Sun (in fact at each second its distance is known to 10 significant figures).

New Horizons will fly past Pluto on 14 July 2015, though it will be scrutinising the Pluto system about six months before that with LORRI (long range imager) and Ralph (high resolution colour mapper).

At 70 days before flyby they will obtain images of Pluto with a spatial resolution equal to that of the Hubble Space Telescope, bettering the HST ever more as the distance decreases towards closest approach. Figure 8.5 is an artist's impression of *New Horizons* approaching Pluto (lower left), with Charon, Nix and Hydra in view. The present plan is to fly past Pluto at a closest approach of about 10 000 km, at a speed of 13.78 km per second. Its closest approach to Charon is planned to be at about 27 000 km.

New Horizons will then spend the following five years traversing the classical E-K belt, doubtless encountering other KBOs. These will need to be no further away than about 55 AU from the Earth, which is about the limit of the communications range, and with the power output from the radioisotope thermoelectric generator declining, observations will be increasingly hindered as time goes by. Also, the KBOs will need to be within a cone with its axis along *New Horizons'* unadjustable path with a cone angle of only 0.2°, because insufficient fuel for manoeuvring would be available for large trajectory changes. Unfortunately this rules out Eris, the largest known KBO (Table 6.1). The spacecraft will, however, have the potential to discover KBOs in the direction of the Milky Way (Figure 1.2), a region of sky so peppered with stars that it is difficult to find KBOs there from the Earth.

Future missions?

No further missions to Pluto are yet planned. Will spacecraft one day land on Pluto, possibly *en route* to the stars? If humans go to Pluto, what will it be like? The final chapter addresses these questions.

9   Pluto: Gateway to beyond?

What would it be like to stand on Pluto: what would we see, what would we feel? Would Pluto be useful as a launch pad for spacecraft to go to other Kuiper belt objects, to the Oort cloud and even to the stars?

9.1 TO STAND ON PLUTO

The sky

The distances from Pluto to the stars are so very much greater than from Pluto to the Earth, that the same constellations will appear in Pluto's sky as in the Earth's sky, and the same Milky Way, all with the same relative brightnesses. The retrograde rotation of Pluto means that the stars will rise in the West and set in the East. The solar day on Pluto, as on the Earth, is the time between successive noons (at noon the Sun is at its maximum altitude). On Earth this interval is one (solar) day. Pluto's axial rotation period is longer than that of the Earth, and therefore the solar day is longer, 6.387 Earth days. The Sun, planets, and stars thus move considerably more slowly across Pluto's sky than across our skies.

The rotation axis on Earth is directed at a point in the northerly sky near the fairly bright star Polaris (the Pole Star), which is a little under 1° from the exact point around which the sky appears to rotate. On Pluto the corresponding point is about 15° East from the bright star Altair. Also, the inclination of the rotation axis is over twice that of the Earth, 57.5° versus 23.4°. This means that the seasonal swing in the altitude of the noon Sun is considerably greater on Pluto (Figure 9.1).

One striking difference in the sky of Pluto, compared with that of the Earth, is that the sky is always black, and clear down to the

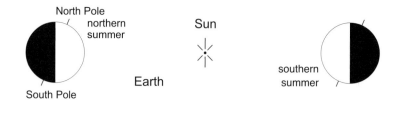

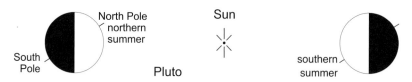

FIGURE 9.1 Illustrating the inclinations of the rotation axes of Pluto and the Earth (not to scale!).

horizon. On Earth, the cloud-free sky is blue. This is because of the scattering of sunlight by molecules in the atmosphere, scattering in all directions. Therefore, a point in the atmosphere in a direction well away from the Sun will scatter some light towards you. The light has a bluish tint because molecules are more efficient scatterers of blue light than of the longer visible wavelengths. Pluto, you have seen, has a very tenuous atmosphere, and the scattering is so weak that it is far too faint for the human eye to see.

When the Sun is in the sky on Pluto, it is by far the brightest object (Figure 9.2). It does, however, bathe Pluto's surface much more feebly than does the Sun from a clear sky on Earth. This is, of course, because Pluto is much further from the Sun than is the Earth, about 31.5 times further at the present time. You might think that this makes the illumination about 31.5 times less. It's a lot worse than this, because sunlight spreads out over a sphere of increasing radius as it travels away from the Sun. The surface area of this sphere increases as the square of its radius, so the illumination on a Sun-facing surface on Pluto is about 31.5 × 31.5 less than on such a surface on Earth,

FIGURE 9.2 An artist's impression of the view from *New Horizons* at Pluto. The Sun is the brilliant light above the crescent Charon. Note Pluto's tenuous atmosphere. (NASA/JHUAPL/SwRI) (See plate section for colour version.)

i.e. about 1000 times less. When allowance is made for the sunlight scattered back to space by a clear sky on Earth this factor becomes somewhat smaller, about 800 times less on Pluto than when the Sun is overhead on the Earth, and increasingly smaller as the altitude of the Sun decreases and the path of sunlight through the Earth's atmosphere increases. On Pluto, during daylight, it would be easy to read and to find one's way around without artificial light. By comparison, even the full Moon's illumination on the Earth is barely sufficient for reading. Also, on Pluto, our colour vision would be activated, unlike the case of moonlight on Earth.

Though the Sun is far and away the brightest object in Pluto's sky, it is far from being the largest object. The Sun's angular diameter as seen from Pluto at its present distance of 31.5 AU from the Sun is about 1.0 minutes of arc (about 0.017°). From the Earth, about 31.5 times closer, the angular diameter of the Sun is about 31.5 times larger, about 30 minutes of arc (about 0.50°). By contrast, the angular diameter of Pluto's sister planet/large satellite Charon is a whopping 3.5° across in Pluto's sky. Our Moon is just 0.50° across in our sky. But even though Charon's albedo is nearly three times that of the Moon, it is so much further from the Sun that it would have a rather dim surface.

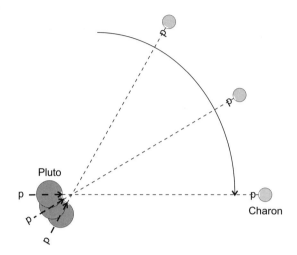

FIGURE 9.3 Pluto and Charon face each other. The points marked p on Pluto and Charon are particular points on each body's surface, in this case the point directly facing the other body.

Recall from Section 4.1 that Charon orbits in Pluto's equatorial plane, its orbital period being the same as the rotation period of Pluto, 6.387 days, and in the same direction. This means that, from any point on the surface of Pluto, Charon is fixed in the sky, neither rising nor setting. It follows from this that from one hemisphere of Pluto, Charon is never visible. The complete picture is that Pluto and Charon face each other, and so Pluto is fixed in Charon's sky too (Figure 9.3), and from one hemisphere of Charon Pluto is never visible. In the case of the Earth and the Moon, whereas the Moon faces the Earth, the Earth does not face the Moon. This is because the Earth is so much more massive than the Moon that our satellite has not sufficiently slowed the Earth's rotation (through tidal forces), though in the distant future it will have done so.

The tiny satellites Nix and Hydra will be fairly bright in Pluto's sky. Whether they will be seen as other than starlike depends on their diameters, which are uncertain. If their angular diameters exceed about 2 minutes of arc their phases (full, gibbous, half, crescent) will be just discernable to an astronaut on Pluto with keen vision (Figure 9.4).

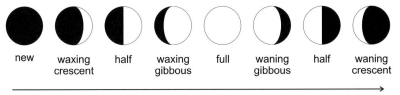

| new | waxing crescent | half | waxing gibbous | full | waning gibbous | half | waning crescent |

time increasing

FIGURE 9.4 The phases of a satellite seen from its planet.

The diameters given in Table 4.2 are very uncertain but, if accurate, Nix would have an angular diameter in Pluto's sky of 6.2 minutes of arc, and Hydra 3.8 minutes of arc. Their orbital periods around Pluto are considerably longer than 6.387 days, so they will rise and set and be visible from all over Pluto (though, because they orbit in Pluto's equatorial plane they will be on the horizon as seen from Pluto's poles).

No KBOs approach Pluto closely enough to be visible from Pluto with the unaided eye.

Planets in Pluto's sky

What about the planets? Would the Earth be visible? Would Jupiter? And what about Pluto's neighbours, Uranus and Neptune?

The brightness of any planet depends on its distance from Pluto, on its phase (Figure 9.4) and on its albedo. The distance of Pluto from the Sun in the period 2009–2020 is 31.5 AU (to three significant figures), and the conclusions below are based on this value.

The Earth, as seen from Pluto, will go through all its phases once per Earth orbit, i.e. once a year, just as our Moon goes through its phase once a month. (In fact the time will be just over a year because of the slow progress of Pluto around its orbit.) Full Earth will be when the Earth is on the far side of the Sun, but because of Pluto's orbital inclination the full Earth would just be visible, provided that the light from the Sun was blocked out. Indeed, the Earth is never more than 1.8° from the Sun, so such blocking would always be necessary for us to be able to see the Earth. Our planet would be at its brightest near to full, but it would be nowhere near as bright as the brightest planets as we see them from the Earth, but more like a rather faint star. Optical aid would be needed to discern a disc (or any other phase) – it would be a point of light to our eyes on Pluto.

Jupiter, as seen from Pluto, will go through its phases in just over the orbital period of Jupiter, which is 11.86 years. As for the Earth, Jupiter is at its brightest when it is not far from full, and will be about as bright as the brighter stars in the sky. Half Jupiter, which in Pluto's skies would still be pretty bright, would be a little over 9° from the Sun, so blocking of sunlight is easier than for the Earth, and with the Sun just below the horizon and Pluto's thin atmosphere, Jupiter would be easily seen, though as a point of light.

Saturn, at a given phase as seen from Pluto, would be dimmer than Jupiter but would be readily visible to the unaided eye. Venus, phase for phase, would be roughly as faint as the Earth. Mars at its brightest would be just beyond the threshold of visibility of normal eyes, and Mercury even further beyond it.

In the cases of Uranus and Neptune, their orbital periods are so long, 84 years and 166 years respectively, that even though I've obtained their brightnesses and phase for 2009, the conclusions will be good for several years after that. Uranus is currently (mid-2009) 35.2 AU from Pluto, and presents a thick gibbous phase. Its brightness is considerably less than half Jupiter – only the most sensitive human eyes could detect it without optical aid; most of us would not be able to see it. Neptune, 28.0 AU from Pluto, also presents a thick gibbous phase, and is slightly fainter than Uranus, just beyond visibility for even the best eyes. Both of these planets would also be points of light to our eyes on Pluto.

From the Earth's surface, stars and planets are dimmed by absorption in the Earth's atmosphere. For an object overhead the dimming is slight, but near the horizon, where the path through the atmosphere is much greater than at the zenith, the dimming is huge (about 40 times greater air mass is traversed near the horizon than at the zenith). You are familiar with this with the rising or setting Sun, which can be viewed comfortably with no eye protection, whereas at higher altitudes the Sun will damage our eyes if viewed without substantial protection (sunglasses won't do!). On Pluto, with almost no atmosphere, there is negligible dimming even at the horizon; you could watch the stars and planets as soon as they rise, up to when they set, with no perceptible change in their brightness.

Another difference from the Earth is that the stars don't twinkle in Pluto's skies. Twinkling in Earth skies is a consequence of turbulence in the Earth's atmosphere. It does not affect planets because their angular diameters, though tiny, are far greater than those of even the nearest stars, which are effectively points of light; the twinkles caused by the various points on a planet's disc cancel out. Therefore, on Pluto, stars and planets shine down steadily, and furthermore there are never any clouds to block our view.

The surface

At the surface the lack of an appreciable atmosphere means that the plutonauts will have to wear spacesuits, and be equipped with heating sufficient to keep a person warm in the frigid surface temperatures, just 40 K or so on the surface of Pluto's Sun-facing hemisphere, and far colder on the hemisphere on which the Sun has set.

A plutonaut will at once be aware of Pluto's low gravity, about half that on the Moon's surface and only about one twelfth of that on the Earth's surface. If the spacesuit were of negligible mass this means that a fairly fit plutonaut would be able to jump upwards, standing upright, 12 metres above the surface of Pluto, but only one metre on Earth. If the spacesuit is the same mass as the plutonaut this figure is halved to 6 metres. Also, whereas a fall from a height of two metres is not fatal on Earth (provided that the person falls feet first and is not extremely frail), the corresponding height on Pluto is 24 metres.

The surface itself will be icy, not so much sheet ice as finely divided, rather like frost. Most of the ice is N_2, intimately mixed with small quantities of CH_4 ice and CO ice, but here and there are patches rich in CH_4 ice and CO ice (Section 5.2). These volatile ices are surely underlain by the less volatile H_2O ice. Some regions will have had their ices darkened by solar UV radiation. Elsewhere a sublimation-deposition cycle will have refreshed the frost. There could also be refreshment from N_2-emitting geysers, as on Triton.

Impact craters are surely present on Pluto; no solid surface in the Solar System is free from them, and there were plenty of impacting bodies early in Pluto's history, and there is no shortage of potential impactors today. The larger impacts would have excavated H_2O ice, both as solids and as vapour. Smaller impacts would vaporize the near surface ices. The degree to which impact craters are obliterated depends on the geological activity, but even if geysers exist, it is unlikely that they could have obliterated the larger craters.

The topography on Pluto, aside from impact craters, is unlikely to be smooth. The smoothest surface in the Solar System is that of Europa, one of Jupiter's four large satellites, but this is because there is a global ocean of liquid H_2O beneath its thin, icy carapace. Pluto is expected to have a surface that suffered upheaval through tidal forces caused by Charon's gravity, until Pluto and Charon evolved to face each other. Also, geological activity in Pluto, additional to any geysers, is certainly possible, both now and in the distant past, as outlined in Section 5.4. There could certainly be mountain ranges, cracks, rift valleys, and even cryovolcanoes.

Figure 9.5 shows an artist's impression of what it would be like to stand on Pluto's surface. Note how grey the icy deposits are, due, at least in part, to the low light levels. I think the relief on Pluto is unlikely to be quite as jagged as that shown.

Many of the questions I've raised about Pluto in this book can be solved by spacecraft such as *New Horizons*, and by robotic landers. So, why would we want to go to Pluto? Why risk human life in such an enterprise? One reason is that geologists and geophysicists on the surface of a planet can achieve considerably more than even highly sophisticated robots, though sheer adaptability. The penalty is the enormous rise in mass and therefore cost of a space mission that has to keep fragile humans alive for many years. Aside from this

FIGURE 9.5 An artist's impression of the view from the surface of Pluto, with a thin crescent Charon to the left and the Sun the brilliant object to the right. Low on the horizon Pluto's thin atmosphere is visible because of our long line of sight through it; it has been exaggerated here. (David Seal, NASA JPL-Caltech)

scientific justification is the human drive to explore, a drive that permeates almost all cultures, and has done so throughout human history; humans did not stay in Africa but spread to almost the whole globe, and though the Apollo landings on the Moon are now 40 years in the past, humans in Earth orbit are now commonplace.

The final reason I give you for which humans might go to Pluto is to build and launch spacecraft to explore the outer Solar System, even as far as the Oort cloud and further to the stars. What advantages, if any, could Pluto offer as a launch platform for the exploration of such far-away regions?

9.2 PLUTO: A LAUNCH PLATFORM?

Let's start by considering a space vehicle – a spacecraft plus its propulsion system – assembled on the Earth and then launched towards the stars, exploring the E-K belt and the Oort cloud en route (Figure 9.6). Energy is required to lift the space vehicle, initially to rise against the

FIGURE 9.6 A space vehicle at the start of its journey to the stars, via the E-K belt and the Oort cloud. This is a NASA design study, called Project Orion, of a craft powered by nuclear fission. (NASA)

pull of the Earth's gravity, then against the pull of the Sun's gravity. If the propulsion system is powered by chemical fuels then these will be heavily depleted early in the flight. The crucial point is that once the spacecraft has reached Pluto, most of the energy needed to overcome the Sun's gravity will have been expended, and the craft would have enough speed to travel beyond Pluto and reach its target in a reasonable time.

If a space vehicle were to be launched from Pluto, then clearly far less energy will be required to reach the E-K belt or the Oort cloud than for a launch from the Earth. But you've doubtless realised that we have to get the spacecraft to Pluto first. The energy required to do this, plus the energy of a re-launch from Pluto, would exceed the energy required *without* any landing on Pluto. It would only make sense to land on Pluto if the propulsion system could be refuelled from materials naturally occurring there.

Chemical propulsion requires the mixture of two substances that react with each other and release a lot of energy in the form of very hot gases that are expelled at high speed, thus propelling the space vehicle in the opposite direction. Reactions that release energy are called exothermic. A lot of energy per unit mass is released if the reactants are a substance rich in oxygen and another rich in hydrogen. Currently, the following reactants are the most commonly used:

- Liquid hydrogen (H_2) burning in liquid oxygen (O_2)
- Kerosene, a liquid also called paraffin (a mixture of hydrocarbons), burning in liquid O_2
- Hydrazine (N_2H_4) burning in N_2O_4 – both are liquids.

Unfortunately, these are either not available on Pluto, or are very likely to be available only in tiny amounts. It might be feasible to manufacture reactants from naturally occurring substances, such as H_2O ice, which can be split into H_2 and O_2 by solar UV radiation, but I very much doubt that useful quantities could be produced in a reasonable time.

A propulsion system that does not use chemical fuels, such as radiation pressure from the Sun, or the highly exothermic nuclear fusion (as in the Sun), would certainly not benefit from a landing on Pluto, unless H_2O ice on Pluto was needed to replenish the hydrogen needed for fusion.

The most exothermic reactions, far more even than fusion reactions, are when matter reacts with antimatter. Antimatter, like matter, consists of atomic particles, but has certain properties that are opposite to those of matter. Thus, whereas an anti-proton has the same mass as a proton, it has the opposite electric charge – a proton is positively charged and an anti-proton is negatively charged. Likewise, an anti-electron, called a positron, is positively charged, whereas an electron is negatively charged. Antimatter would have to be produced on Earth in very sophisticated machines. The big problem is storage; when matter and antimatter meet, they both disappear in a huge explosion. But if, in the distant future, a technology was developed for a controlled reaction, then the stars would be within reach. However, the energy required to reach the outer Solar System is truly tiny compared with the energy required to reach even our nearest star, Proxima Centauri, 4.22 light years away, in a reasonable time. At 1% of the speed of light, the journey time would be 422 years! With antimatter drive this could, perhaps, be reduced to a few tens of years. In any case, Pluto, devoid of antimatter, would offer no advantage as a re-launch platform.

One way in which a landing on Pluto would enable enhanced exploration of the far reaches of the Solar System, would be for a mission, with or without humans, to erect a large robotic telescope, which would continue to be operated remotely even after the humans from any manned landing had returned to Earth.

Coda

So there we are. I hope that you have enjoyed my story of how we discovered Uranus, Neptune, Pluto and the host of other bodies in the outer Solar System. I hope also that you can appreciate how these distant bodies have enhanced our understanding of the formation and evolution of the Solar System. My discussions of Pluto and the E-K belt led me to explain techniques and concepts in astronomy of wide applicability, and the classification of Pluto is an excellent case study of the role and difficulties of classification in science. The exploration of the outer Solar System continues, and we look forward to the flyby of the Pluto system by *New Horizons* in 2015, and beyond into the E-K belt, to give a big boost to our knowledge and understanding of these small worlds, and of our Solar System.

Glossary

This short glossary is confined to objects, concepts, instruments, missions and methods that are specific to this book and central to it. The many other objects, etc., that appear in this book (including those in this glossary) are listed in the index, from where you can find the page(s) in the text where they are defined or described, sometimes more fully than here. However this glossary, in many instances, pulls together material distributed around several locations in the main text.

Though you do not need to do so in relation to this book, you might find a dictionary of astronomy useful for a quick look-up of basic astronomical terms and concepts. One such is the *Cambridge Illustrated Dictionary of Astronomy* (details at http://www.cambridge.org/9780521823647), which is available for purchase in printed and electronic form. Alternatively, several publishers have a dictionary of astronomy in their lists, for example Collins. Any large bookshop should stock at least one dictionary of astronomy.

Quick look-ups are also provided via the internet, for example via Google, where you can also find information beyond that which I've presented to you in this book. Treat all information you find via Google with care. Make sure that an article has a list of references and that it is up to date. Fortunately, astronomy entries are less prone to misleading or incorrect entries than is the case in some other subjects.

Albedo An albedo is a measure of the reflectivity of a surface. In astronomy the most important albedos are the geometric albedo and the Bond albedo. The *geometric albedo* is the amount of solar radiation a body reflects towards us compared with the amount we would receive from a flat Lambertian surface, which is a flat surface that reflects 100% of the radiation incident upon it, and is perfectly diffuse (the opposite of a mirror). The *Bond albedo* is the fraction of the solar radiation intercepted by a body that is reflected back to space. It differs from the geometric albedo, which is the radiation reflected *towards the observer* compared with that from a standard surface (a Lambertian surface). In the text I have abbreviated geometric albedo to albedo.

Astronomical unit (AU) This used to be defined as the semimajor axis of the Earth's orbit, but because this varies very slightly, the AU has now been fixed as 149.5978715 million kilometres.

Blink comparator A device used to compare two images of the sky by switching between them. It is used to find bodies that are moving with respect to the background of the distant stars.

Centaurs Small bodies in unstable orbits between Jupiter and Neptune. They are a transient population between the scattered disc objects in the E-K belt and (nearly all of) the short-period comets.

Charon Pluto's large satellite. It is icy-rocky in composition and has a mass 11.7% that of Pluto, sufficient that the centre of mass of Pluto and Charon lies outside Pluto. No atmosphere has been detected. Pluto and Charon always present the same faces to each other.

Classification in science It provides an economy of description, a tool for structuring knowledge, and can also lead to deeper understanding.

Comets There are two main classes of comet, the short-period comets and the long-period comets. The somewhat arbitrary dividing line is that short-period comets are those that have orbital periods up to 200 years, and long-period comets have orbital periods greater than 200 years. More fundamental distinctions are:

1 the short-period comets are predominantly in orbits of no more than modest inclination and modest eccentricity, whereas the long-period comets display the full range of orbital inclinations and have high orbital eccentricities

2 the source of the great majority of the short-period comets is the scattered disc, whereas the source of all the others is the Oort cloud.

Dwarf planets At the 2006 triennial meeting of the International Astronomical Union, held in Prague, the definition of a planet was considered. One outcome was the definition of a class called dwarf planets. The defining characteristics are a celestial body that:

1 is in orbit around the Sun
2 has sufficient mass for its self-gravity to overcome rigid body forces so that it assumes a hydrostatic equilibrium (nearly round) shape
3 has *not* cleared the neighbourhood around its orbit
4 is not a satellite.

At present, the dwarf planets comprise Ceres (the largest asteroid), and four of the largest Kuiper belt objects, Pluto (the largest plutino), Eris, Makemake and Haumea.

Edgeworth-Kuiper belt (E-K belt) Icy-rocky bodies in orbits of fairly low inclination, the semimajor axes of the great majority of them lying in the range from a little over 40 AU to 47.7 AU. The belt comprises the plutinos, the classical objects and the scattered disc objects.

Icy materials Substances that, when solid, are icy in appearance, and melt to become liquid, or sublime to become gas, at comparatively low temperatures.

Kuiper belt objects (KBOs) Members of the Edgeworth-Kuiper belt. There are three distinct sorts.

1 The classical KBOs account for about two thirds of the 2000 or so known KBOs, and have semimajor axes ranging from a little over 40 AU to 47.7 AU. The majority are in orbits of no more than modest eccentricity and inclination.
2 The 200 or so known Plutinos occupy the 3:2 mean motion resonance with Neptune, at 39.4 AU. Most have orbital

eccentricities the order of that of Pluto, and orbital inclinations that range from around that of Pluto downwards. They account for about 10% of the known KBOs.

3 The scattered disc objects (SDOs) constitute a rather sparse population occupying unstable orbits, typically with much greater orbital eccentricities and inclinations than found among the classical KBOs and the Plutinos. They are the likely source of the Centaurs and hence most of the short-period comets.

Mean motion resonance If a body orbits around the Sun with a certain period, and another body does so with its own period, and the ratio of the periods can be expressed by small digits, then the orbits are in a mean motion resonance. For example, for every three orbits of Neptune, Pluto orbits twice. This means that the ratio of the orbital period of Pluto to that of Neptune is 3:2, which is called the three-to-two mean motion resonance. This particular resonance is stable, i.e. the orbital elements do not wander.

New Horizons The first spacecraft to be launched towards Pluto and the E-K belt beyond. It was launched from Cape Canaveral on 19 January 2006, has flown past Jupiter at fairly close range, and will fly past Pluto on 14 July 2015.

Occultation When a body passes between an observer and a smaller body such that the light from the smaller body is completely blocked, an occultation (hiding) is said to have occurred.

Oort cloud The outer component is a thick spherical shell of icy-rocky bodies surrounding the Solar System, extending from about 1000 AU (perhaps 10 000) to about 100 000 AU. The inner component is more belt-like, and extends inwards from the outer Oort cloud towards the E-K belt. It is estimated that the Oort cloud contains somewhere between about a million million and ten million million bodies with sizes greater than a kilometre. They are the source of the long-period comets.

Planet At the 2006 triennial meeting of the International Astronomical Union, held in Prague, the definition of a planet was considered. Their defining characteristics are a celestial body that:

1 is in orbit around the Sun

2 has sufficient mass for its self-gravity to overcome rigid body forces so that it assumes a hydrostatic equilibrium (nearly round) shape

3 has cleared the neighbourhood around its orbit.

In the Solar System the class planet presently comprises Mercury, Venus, Earth, Mars, Jupiter, Saturn, Uranus and Neptune.

Planet X This was the name that Percival Lowell gave to the planet beyond Neptune held responsible for the discrepant motions of Uranus. In the years up to 1914 he produced several masses, orbits, and positions of his Planet X, and he conducted several searches, without success. When Clyde Tombaugh searched (successfully) for a planet beyond Neptune (and found Pluto), he made no use of Lowell's predictions.

Pluto The planet beyond Neptune discovered by Clyde Tombaugh in 1930, after a systematic search (using a blink comparator). This small icy-rocky world, 0.218% the mass of the Earth, cannot account for the discrepant motions of Uranus. Indeed, in 1993 *Voyager 2* showed that Neptune's mass was 0.5% larger than previously thought, which accounted for Uranus's apparent discrepant motions. Pluto has a very tenuous atmosphere. It has been classified as a dwarf planet, and is the largest known plutino.

Rocky materials These materials are denser than icy materials and have higher melting points. Common rocky materials are silicates. Others include metal oxides. Iron, and its alloy with nickel, Fe-Ni, is sometimes included in rocky materials, for the sake of brevity.

Small Solar System bodies At the 2006 triennial meeting of the International Astronomical Union, held in Prague, small Solar System bodies were defined as everything orbiting the Sun, except satellites, planets, their satellites, and dwarf planets. These currently include the asteroids (except for Ceres, which is a dwarf planet), KBOs that are not dwarf planets, comets, and other small bodies.

Transit When a body passes between the observer and another body with a larger size, a transit is said to have occurred.

Triton Neptune's largest satellite, has a diameter 17% greater than that of Pluto, and also a similar mean density, indicating a broadly similar composition to Pluto. Its orbit is retrograde, which can be explained by capture from the Pluto region.

Trojans Small bodies in stable orbits around a point 60° ahead of a planet and 60° behind it (which are two of the five Lagrangian points). They were first discovered in Jupiter's orbit, where about 3000 Trojans are known. Neptune is known to have a dozen or so, a number that will surely rise.

Further reading and other resources

SOME OF THE KEY PAPERS IN THE SCIENTIFIC LITERATURE

Journal titles are in *italics*, volume or issue numbers are in **bold**, and the first and last page numbers are given. If there are more than three authors the first author is given, followed by *et al*. Most of these papers require a basic knowledge of astronomy and planetary science. The papers are in year order, starting with the most recent. A representative selection is given with no attempt at being comprehensive.

THE DISCOVERY OF URANUS, NEPTUNE AND PLUTO
(See also Books)

Uranus

Account of a Comet, By Mr Herschel, FRS; Communicated by Dr Watson, Jun. of Bath, FRS (1781) W Herschel, *Philosophical Transactions of the Royal Society of London* **71** 492–501

General Notes and Discoveries (1781) ed. J E Bode, *Berliner Astronomisch Jahrbuch*, 210–220

Neptune

J C Adams, Cambridge and Neptune (1996) D W Hughes, *Notes and Records of the Royal Society* **50** 245–248

Neptune's Discovery 150 Years Later (1996) W Sheehan and R Baum, *Astronomy* **24**(9) 42–49

Private Research and Public Duty: George Biddell Airy and the Search for Neptune (1988) A Chapman, *Journal for the History of Astronomy* **19**(2) 121–139

Le Centenaire de la Découverte de Neptune (1946) A Danjon, *Ciel et Terre* **62** 369–383

Account of Some Circumstances Historically Connected with the Discovery of the Planet Exterior to Neptune (1847) G B Airy, *Memoirs of the Royal Astronomical Society* **16** 385–414

Pluto

W H Pickering's Planetary Predictions and the Discovery of Pluto (1976) W G Hoyt, *Isis* **67 (4)** 551–564

The Search for a Planet Beyond Neptune (1964) M Grosser, *Isis* **55 (2)** 163–183

The Search for the Ninth Planet, Pluto (1946) C W Tombaugh, *Astronomical Society of the Pacific Leaflets* **5** 73–80

The Discovery of Pluto (1931) A C D Crommelin (quoting Nicholson and Mayall), *Monthly Notices of the Royal Astronomical Society* **91** 380–385

THE CLASSIFICATION OF PLUTO

(See also Books and Internet links)

What is a Planet? (2007) Stephen Soter, *Scientific American* 2007 (January) 34–41

The controversial change of Pluto's classification to being a dwarf planet has flaws, but on the whole encapsulates scientific principles.

What happened to Pluto? (2006) Owen Gingerich, *Sky and Telescope* 2006 (November) 34–39

An account of the deliberations at the IAU General Assembly in August 2006, in Prague, that resulted in the class dwarf planet being defined, and Pluto being placed in it.

PLUTO AND ITS SATELLITES

Note that many of these are rather advanced.

Pluto's Lower Atmospheric Structure and Methane Abundance from High-resolution Spectroscopy and Stellar Occultations

(2009) E Lellouch *et al.*, *Astronomy and Astrophysics* **495** L17–L21

One of the 8-metre telescopes that constitute ESO's VLT in Chile was used.

Masses of Nix and Hydra (2008) D J Tholen *et al.*, *Astronomical Journal* **135** 777–784

A four-body solution to the orbits of Pluto, Charon, Nix and Hydra leads not just to the masses of Nix and Hydra, but also to revised masses of Pluto and Charon.

Waves in Pluto's Upper Atmosphere (2008) M J Person *et al.*, *Astronomical Journal* **136** 1510–1518

Results from the March 2007 stellar occultation by Pluto are reported. Amongst these it is confirmed that the significant increase in atmospheric pressure between 1988 and 2002 has ceased.

On the Origin of Pluto's Minor Moons, Nix and Hydra (2008) Y Lithwick and Y Wu, *arXiv:astro-ph* 0802.2951

They conclude that the only model that works for both Nix and Hydra together is that they were captured debris from the nebular disc.

Structure and Evolution of Kuiper Belt Objects and Dwarf Planets (2008) W B McKinnon *et al.*, in *The Solar System Beyond Neptune*, University of Arizona Press, eds. M A Barucci, H Boehnhardt and D P Cruikshank, 213–241

Pluto and Charon are included as dwarf planets.

Changes in Pluto's Atmosphere 1988–2006 (2007) J L Elliot *et al.*, *Astronomical Journal* **134** 1–13

Results from the stellar occultations by Pluto in 1988, 2002 and 2006 are reported. The rise in pressure from 1988 to 2002 has ceased. A revised radius of Pluto is reported.

Near Infrared Spectroscopy of Charon: Possible Evidence for Cryovolcanism on Kuiper Belt Objects (2007) J C Cook *et al.*, *Astrophysical Journal* **663** 1406–1419

The title is self explanatory.

Charon's Radius and Density from the Combined Data Sets of the 2005 July 11 Occultation (2006) M J Person *et al.*, *Astronomical Journal* **132** 1575–1580

The title is self-explanatory.

Discovery of Two New Satellites of Pluto (2006) H A Weaver *et al.*, *Nature* **439** 943–945

The discovery in 2005, in HST images, of what, in 2006, were named Nix and Hydra.

Sub-arcsecond Scale Imaging of the Pluto/Charon System at 1.4 mm (2005) M A Gurwell and B J Butler, *Bulletin of the American Astronomical Society* **37** 743–?

Using the Submillimeter Array (SMA) average temperatures are obtained for the surfaces of Pluto and Charon.

A Giant Impact Origin of Pluto-Charon (2005) R M Canup, *Science* **307** 546–550

The title is self-explanatory.

Pluto and Charon: Formation, Seasons, Composition (2002) M E Brown, *Annual Review of Earth and Planetary Science* **30** 307–345

The title is self-explanatory.

A Two-Color Map of Pluto's Sub-Charon Hemisphere (2001) E F Young, R P Binzel and K Crane, *Astronomical Journal* **121** 552–561

Maps from the mutual events 1987–1989.

Distribution and Evolution of CH_4, N_2, and CO ices on Pluto's surface: 1995 to 1998 (2001) W M Grundy and M W Buie, *Icarus* **153** 248–263

Near IR observations are used to constrain the longitudinal distributions of the three ices.

Composition, Internal Structure, and Thermal Evolution of Pluto and Charon (1997) W B McKinnon, D P Simonelli and G Schubert, in *Pluto and Charon*, University of Arizona Press, eds. S A Stern and D J Tholen, 295–343

The title is self-explanatory.

Surfaces of Pluto and Charon (1997) D P Cruikshank *et al.*, in *Pluto and Charon*, University of Arizona Press, eds. S A Stern and D J Tholen, 221–267
  A comprehensive review up to 1996.
Detection of Gaseous Methane on Pluto (1997) L A Young *et al.*, *Icarus* **127** 258–262
  The first detection of gaseous methane in Pluto's atmosphere, from near IR observations.
Atmospheric Structure and Composition: Pluto and Charon (1997) R V Yelle and J L Elliot, in *Pluto and Charon*, University of Arizona Press, eds. S A Stern and D J Tholen, 347–390
  A comprehensive review up to 1996.
Pluto's Heliocentric Orbit (1997) R Malhotra and J G Williams, in *Pluto and Charon*, University of Arizona Press, eds. S A Stern and D J Tholen, 127–157
  The dynamical evolution of Pluto's peculiar orbit around the Sun.
The Origin of Pluto's Orbit: Implications for the Solar System Beyond Neptune (1995) R Malhotra, *Astronomical Journal* **110** 420–429
  The outward migration of Neptune plays the crucial role.
Albedo Maps of Pluto and Charon – Initial Mutual Events Results (1992) M W Buie, D J Tholen and K Horne, *Icarus* **97** 211–227
  From observed lightcurves, albedo surface maps are obtained for the surfaces of Pluto and Charon.
Pluto's Atmosphere (1989) J L Elliot *et al.*, *Icarus* **77** 148–170
  Airborne CCD photometer observations of Pluto's 9 June 1988 stellar occultation yield an occultation lightcurve, which reveals an upper atmosphere overlying an extinction layer.
A Two-spot Albedo Model for the Surface of Pluto (1988) R L Marcialis, *Astronomical Journal* **95** 941–947
  An albedo map of Pluto showing two dark circular spots in the equatorial region embedded in a less dark band centred on the equator.

Why is Pluto Bright? Implications of the Albedo and Lightcurve
   Behaviour of Pluto (1988) S A Stern, L M Trafton and G R Glad-
   stone, *Icarus* **75** 485–498
   Includes lightcurves of Pluto.
The Separate Spectra of Pluto and its Satellite Charon (1988) U Fink
   and M A Disanti, *Astronomical Journal* **95** 229–236
   From the occultation of Charon by Pluto on 3 March 1987, it
   is concluded that Charon has no discernable methane atmo-
   sphere and, unlike Pluto, has no more than very sparse methane
   frost.
Water Frost on Charon (1987) M W Buie *et al.*, *Nature* **329** (8 October)
   522–523
   Spectra taken just before and during a total eclipse of Charon by
   Pluto reveal the spectral signature of water ice on Charon, but no
   evidence for methane or ammonia frosts.
The Detection of Eclipses in the Pluto-Charon System (1985) R P
   Binzel *et al.*, *Science* **228** (7 June) 1193–1195
   The first eclipses between Pluto and its satellite Charon were
   detected in January and February 1985, allowing greatly improved
   determinations of the diameters of the planet and satellite, the
   surface albedo distribution on one hemisphere of Pluto, its mass,
   and the orbit of Charon.
From the Ridiculous to the Sublime: the Pending Disappearance of
   Pluto (1980) A J Dessler and C T Russell, *EOS* **61** 690
   Pluto's decreasing mass in the decades since its discovery.
The Satellites of Neptune and the Origin of Pluto (1979) R S Harrington and T van Flandern, *Icarus* **39** 131–136
   The origin of Pluto and the peculiar satellite system of Neptune,
   Triton, as a result of a disruption of a normal system of Neptunian
   satellites by a passing body of significant mass is considered.
The satellite of Pluto (1978) J W Christy and R S Harrington, *Astro-
   nomical Journal* **83**, 1005, 1007–1008
   Astrometric and photographic observations of Pluto are reported
   which indicate that the planet appears to have a faint satellite.
   (This was later confirmed, and it was named Charon.)

Pluto – Evidence for Methane Frost (1976) D P Cruikshank, C B Pilcher and D Morrison, *Science* **194** (19 November) 835–837

From infrared photometry of Pluto from 1.2 to 2.2 micrometres, which includes the diagnostic absorption bands of water and methane frosts, it is concluded that methane frost is probably the dominant reflecting material on the planet's surface.

Spectrophotometry of Pluto (1970) J D Fix, J S Neff and L A Kelsey, *Astronomical Journal* **75** 895–896

The first determination of the geometrical albedo of Pluto, at 21 wavelengths from 0.34 to 0.59 micrometres.

THE EDGEWORTH-KUIPER BELT

(See also Internet links)

Structure and Evolution of Kuiper Belt Objects and Dwarf Planets (2008) W B McKinnon *et al.*, in *The Solar System Beyond Neptune*, University of Arizona Press, eds. M A Barucci, H Boehnhardt and D P Cruikshank, 213–241

The title is self explanatory.

Origin of the Structure of the Kuiper belt during a Dynamical Instability in the Orbits of Uranus and Neptune (2008) H F Levison *et al.*, *Icarus* **196** 258–273

The title is self explanatory.

Satellites of the Largest Kuiper Belt Objects (2006) M E Brown *et al.*, *Astrophysical Journal* **639** L43–L46

A discussion of the special origin of the satellites of three of the four brightest KBOs: Pluto, 2003 EL61 and 2003 UB313. The second brightest KBO after Pluto, 2005 FY6, does not have a satellite with a brightness of more than 1% of the primary.

Near-infrared Surface Properties of the Two Intrinsically Brightest Minor Planets: (90377) Sedna and (90482) Orcus (2005) C A Trujillo *et al.*, *The Astrophysical Journal* **627** 1057–1065

The title is self explanatory.

The Impact of a Close Encounter on the Edgeworth-Kuiper Belt (2004) A C Quillen, D E Trilling and E G Blackman, *arXiv:astro-ph* 0401372

A numerical investigation of the possibility that a close stellar encounter could account for the high inclinations found in the Kuiper belt.

Quaoar and the Edgeworth-Kuiper Belt (2003) D W Hughes, *Astronomy and Geophysics* **44** 3.21–3.22

A brief review of the history of the prediction and discovery of objects beyond Neptune that constitute the Edgeworth-Kuiper belt.

Planetary Science: Out on the Edge (2002) W B McKinnon, *Nature* **418** (11 July) 135–137

A brief account of the E-K belt, and prospects for its exploration.

Kuiper Belt Objects: Relics from the Accretion Disc of the Sun (2002) J X Luu and D C Jewitt, *Annual Reviews of Astronomy and Astrophysics* 63–101

A review of what was known about KBOs in 2002.

The Size and Albedo of the Kuiper-belt Object (20 000) Varuna (2001) D Jewitt, H Aussel and A Evans, *Nature* **411** (24 May) 446–447

A brief account of the discovery of what is now called Varuna, which at that time was the largest KBO after Pluto and Charon, and the measurement of its size and albedo.

2001 KX76 (2001) R L Mills *et al.*, *IAU Circular* **7657**

Hard on the heels of the discovery of Varuna came this discovery of the slightly larger KBO, 2001 KX76, subsequently named Ixion. (It is now thought to be slightly smaller than Varuna.)

The Edgeworth-Kuiper Belt (1996) M T Brück, *Irish Astronomical Journal* **23** 3

Kenneth Edgeworth's prediction of small objects beyond Pluto in 1949, two years before Gerard Kuiper discussed a similar possibility.

Discovery of the Candidate Kuiper Belt Object 1992 QB1 (1993) D C Jewitt and J X Luu, *Nature* **362** (22 April) 730–732

The discovery is reported of a new faint object in the outer Solar System, 1992 QB_1, moving beyond the orbit of Neptune. It is suggested that 1992 QB_1 may represent the first detection of a member of the Kuiper belt.

NEW HORIZONS

(See also Internet links)

The New Horizons Pluto Kuiper Belt Mission: An Overview with Historical Context (2008) S A Stern, *Space Science Reviews* **140** (Issue 1–4) 3–21

The title is self-explanatory.

New Horizons at Jupiter (2007) Eleven articles in a special section of *Science* **318** 215–243

The articles describe the new data obtained about Jupiter, its large satellites, and its rings, from the fly-by of *New Horizons* 28 February 2007.

Baseline Design of New Horizons Mission to Pluto and the Kuiper belt (2006) Y Guo and R W Farquhar, *Acta Astronautica* **58 (10)** 550–559

The title is self-explanatory.

Here is an Alan Stern article from 2002 that describes the mission and science. http://www.scientificamerican.com/article. cfm?id=journey-to-the-farthest-p

This is from *Sky and Telescope*, by Kelly Beatty, but is rather short. http://www.skyandtelescope.com/news/3304671.html

Here is a more substantial article from the *Smithsonian Magazine*. http://www.airspacemag.com/space-exploration/Where-the-Wild-Things-Are.html?c=y=1

Here is a *National Geographic* article. http://news. nationalgeographic.com/news/2005/02/0215_050214_pluto.html

SPACECRAFT PROPULSION SYSTEMS

(See also Internet links)

Engage the Antimatter Drive (2007) B Crystall, *New Scientist* **2620** 62–64

An account of various forms of propulsion systems for interstellar flight.

Warp Speed (2006) Craig Covault, *Aviation Week and Space Technology* **164 (2)** 46–49

This is an article by a leading space journalist.

BOOKS

Books requiring a basic knowledge of astronomy and planetary science are asterisked (*). The books are in year order, staring with the most recent.

The Solar System
This is a small selection.
Discovering the Solar System, 2nd edition, Barrie W Jones, J Wiley & Sons 2007
  ISBN 978 0 470 01831 6
  An introductory textbook on the Solar System.
An Introduction to the Solar System, eds. Neil McBride and Iain Gilmore, The Open University and Cambridge University Press 2003
  ISBN 0 521 54620 6
  An introductory textbook on the Solar System.

Pluto, its satellites and the Edgeworth-Kuiper belt
*The Solar System Beyond Neptune, eds. M A Barucci, H Boehnhardt and D P Cruikshank, University of Arizona Press 2008
  ISBN 0 8165 2755 5
  This is a comprehensive collection of 35 articles on the Solar System beyond Neptune, mainly on the Edgeworth-Kuiper belt but also including a small amount of material on Pluto, Charon and the Oort cloud. The articles are distributed over the following sections: Introduction; Transneptunian object populations; Bulk properties; Physical processes; Formation and evolution; Individualities and peculiarities; Links with other Solar System populations; Boundaries and connections to other stellar systems; Laboratory; and Perspectives.
Is Pluto a Planet?, David A Weintraub, Princeton University Press 2007
  ISBN 978 0 691 13846 6

A historical journey through the Solar System, including the classification of Pluto.

Pluto: a Case Study in Science, J M Pasachoff and M A Seeds, Thomp-
son Brooks/Cole 2007

ISBN 0 495 38405 4

A 16 page pamphlet, mainly about the classification of Pluto.

Pluto and Charon, 2nd edition, Alan Stern and Jacqueline Mitton,
Wiley-VCH 2005

ISBN 3 527 40556 9

A detailed account of the outer Solar System, with an emphasis on
the history of discoveries, but also including our understanding in
the early years of the twenty-first century, and an account of the
New Horizons mission.

Beyond Pluto: Exploring the Outer Limits of the Solar System, John
Davies, Cambridge University Press 2001

ISBN 0 5218 0019 6

Comprehensive coverage of the Edgeworth-Kuiper belt.

Planet Quest: the Epic Discovery of Alien Solar Systems, Ken
Croswell, The Free Press, New York 1997

ISBN 978 0684 83252 4

Includes an account of the discovery of Pluto.

Planets X and Pluto, William G Hoyt, University of Arizona Press
1981

ISBN 0 8165 0664 7

A strong emphasis on the predictions and discovery of Pluto, plus
chapters on the discovery of Uranus and Neptune.

Out of the Darkness: the Planet Pluto, Clyde W Tombaugh and
Patrick Moore, Stackpole Books, Lutterworth Press 1980

ISBN 0 8117 1163 3

A strong emphasis on the predictions and discovery of Pluto.

The Planet Pluto, A J Whyte, Pergamon Press 1980

ISBN 0 0802 4648 6

The discovery of Pluto and a short account of our knowledge of the
planet in 1979.

* *Pluto and Charon*, eds. S A Stern and D J Tholen, University of Arizona Press 1997

ISBN 0 8165 1840 8

This is a comprehensive collection of 20 articles on Pluto and Charon. After two short articles on the discovery of Pluto and Charon, by the discoverers (Clyde Tombaugh and James Christy respectively), the remaining 18 articles are distributed over five sections: Historical perspective; Dynamics; Bulk properties, Surfaces and interiors; Atmospheres; and Perspectives.

MAGAZINES

The monthly publications *Sky and Telescope*, *Astronomy*, *Astronomy Now* and *Sky at Night Magazine* are all aimed at a wide readership and contain articles on the Solar System, plus other areas of astronomy. *Scientific American* (monthly) and *New Scientist* (weekly) are also aimed at a wide readership, and contain occasional articles and short items on the Solar System.

INTERNET LINKS

The IAU resolution made in August 2006, classifying Pluto as a dwarf planet: http://www.iau.org/static/resolutions/Resolution_GA26–5–6.pdf

Discussion of Pluto's classification:

http://www.iau.org/public_press/themes/pluto/

http://news.nationalgeographic.com/news/2006/08/060824-pluto-planet.html

http://blogs.nature.com/news/blog/conference_reports/international_astronomical_union/

http://www.telegraph.co.uk/science/science-news/3349184/Pluto-should-get-back-planet-status-say-astronomers.html

Lists of Trans Neptunian Objects (2009):

http://www.cfa.harvard.edu/iau/lists/TNOs.html

NASA's *New Horizons* website:
   http://pluto.jhuapl.edu/index.php
The position of *New Horizons* in its trajectory, and its distance from
   Pluto, Jupiter, the Earth and the Sun (frequently updated):
   http://www.yaohua2000.org/cgi-bin/New%20Horizons.pl
Proposed interstellar probes using nuclear pulse propulsion: Project
   Orion (including the Soviet version), Project Daedelus, Medusa
   and Project Longshot:
   http://en.wikipedia.org/wiki/Nuclear_pulse_propulsion

Index

Note that unless something substantial is added on later pages, only the first page on which the item appears is given.

CONCEPTS, INSTRUMENTS, MISSIONS, AND METHODS

PEOPLE (not every mention is given)